A
HISTORY
OF
VODKA

A
HISTORY
OF
VODKA

William Pokhlebkin

Translated by Renfrey Clarke

VERSO

London · New York

First published in Russian 1991
This translation first published by Verso 1992
© Interverso 1991
All rights reserved

Verso
UK: 6 Meard Street, London W1V 3HR
USA: 29 West 35th Street, New York, NY 10001–2291

Verso is the imprint of New Left Books

ISBN: 978-0-86091-359-7

British Library Cataloguing in Publication Data
A catalogue record for this book is available from the British Library
Library of Congress Cataloging in Publication Data
A catalogue record for this book is available from the Library of Congress

Typeset in Photina by Goodfellow & Egan Ltd, Cambridge
Printed in Great Britain by Bookcraft (Bath) Ltd

Contents

Translator's Note

There is no single system for transliterating Russian words into English that meets the needs both of scholars and of authors and translators aiming at a mass market. The problem that this presents is magnified in the case of a work such as the present one, directed both at specialists and at general readers who are simply interested in vodka and in learning about the social and historical setting in which it arose.

The solution I have resorted to is to use several systems of transliteration for different purposes. In the notes and for italicized items I have used a simplified variant of the systems commonly employed in British and American libraries. Elsewhere, my practice has been dictated by the need to provide readers who do not know Russian with an approximation of the original sound, while avoiding pedantry and keeping in mind that this is, after all, a book published in English for readers of that language.

In the text, names have been given in their English forms (Catherine the Great; Monastery of the Resurrection) wherever this seemed appropriate.

I would like to thank R.E.F. Smith for his helpful comments on a draft of the translation – while, of course, absolving him from any responsibility for any errors which may remain.

Foreword

How and Why this Book Came to be Written

This text was never intended for publication, and still less was it conceived as a sort of "history of drunkenness" like Pryzhkov's admirable *History of the Russian Tavern*. It was conceived as serious research to clarify a specific, purely historical question: when did the production of vodka in Russia begin, and was this earlier or later than in other countries? In other words, the author's task was to establish whether the production of vodka originated in Russia, or whether, in this most sensitive area of our national identity, someone from Europe or Asia pointed the way.

This question had seldom been raised in the previous two centuries, and it would probably not have arisen in our own time, except for the fact that at the end of 1977 the origins of vodka unexpectedly became an issue of international significance. The year 1977 was, so to speak, the pinnacle of so-called "developed socialism". Apart from the sixtieth anniversary of the October Revolution, that year also saw the adoption of a new constitution and of a series of new laws which were formally directed toward strengthening socialism; though in fact they were drafted and applied in such a clumsy and bureaucratic fashion that they actually harmed it. These developments within the USSR were accompanied by a "thaw", a degree of *détente* in foreign policy. Western leaders could not reject this approach and, despite private misgivings, publicly endorsed it. They could not openly find fault with the Soviet Union's foreign policy line, though the temptation to do so increased as the triumphant festivities surrounding the anniversary continued. So, as was usual in this bygone era of the Cold War, they found a pretext for creating a minor scandal in order to take the gloss off the birthday of Soviet socialism.

In the autumn of 1977, right on the eve of the anniversary celebrations, Soviet brands of vodka including Moscow Special, Stolichnaya, Limonnaya and others began to be subjected to discrimination in foreign markets. Simultaneously, the threat emerged that the Soviet organization Soyuzplodoimport would be stripped of the right to advertise and sell this beverage as vodka, since a number of foreign firms, both European and American, began to claim the exclusive right to use the name "vodka" for their product, on the basis that they had begun producing it earlier than the Soviet firms.

At first the Soviet Ministry of Foreign Trade did not take these claims seriously. The foreign competitor firms based their case on the fact that production of vodka in the USSR began only in 1923, under a decree of 26 August of that year, while they had begun producing it earlier, mainly in the years between 1918 and 1921. During this period many industrialists who had fled from Russia with the Whites set up production of vodka in the West.

From December 1917 the Soviet government had, in fact, banned the production of vodka on the territory of the Russian Soviet Federative Socialist Republic, resuming it only around the end of 1923 and the beginning of 1924. But in juridical and historical terms it was easy to prove that in the first instance, the Soviet government had merely continued a ban on the production of spirituous liquors which had been maintained by the previous Tsarist and Provisional governments throughout the period of the First World War, beginning in 1914; as far as a court would have been concerned, it was simply a matter of an existing temporary prohibition being reaffirmed. Second, it was a simple matter to establish that the state monopoly on vodka production had been inherited from the past, along with the right of the state to shut down or suspend production or renew it as it wished. Consequently, the date 26 August 1923 did not in any sense mark the beginning of vodka production in Russia, and had no bearing on the question of prior right to the use of the name "vodka", since this appellation had arisen not with the beginning of production in the year 1923, but in connection with the earliest production of vodka in Russia. From this it followed that parties claiming that vodka had originated in their countries had to present convincing evidence showing that in some year or other vodka had first been produced on their territory.

Once the question had been put on this historical and juridical plane, all the American and Western European vodka firms – Pierre Smirnoff, Eristov, Keglevich, Gorbachev and others – were forced to drop their claims to the prior invention of Russian vodka, and thereafter could base their advertising only on the supposed special qualities of their products.

The relative ease with which this first assault by the Soviet Union's trading competitors was beaten off encouraged the Ministry of Foreign Trade and its subsidiary organization Soyuzplodoimport to rest on their laurels, with the result that they were completely unprepared when a second attack was launched. This came from the state vodka monopoly of the Polish People's Republic, and not unnaturally the Soviet side saw it as a stab in the back. Only now, more than a decade later, has it become clear to everyone that there was nothing accidental, or even purely commercial, about the Polish action. Anti-Soviet factions had long been gathering strength in Poland. They were closely linked with various reactionary circles in Europe and America, and it was obvious even then that they were acting as agents for the Western "vodka kings" who had failed in their first attack. But in formal terms they claimed to be defending the state interests of the Polish People's Republic. For the Soviet organization Soyuzplodoimport the situation was both unexpected and thoroughly unpleasant; our people were not accustomed to disputes with allies.

Meanwhile the Polish state vodka monopoly was claiming that vodka had been discovered and produced earlier in Poland – that is, on the state territory of the former Kingdom of Poland, the Grand Duchy of Lithuania and the Rech Pospolita, including Greater and Lesser Poland, Masuria, Pomerelia, and the Ukraine including the Cherkassk and Zaporozhe Sech – than in the

Russian lands or on the territories of the Moscow state. As a result of this, they argued, the right to advertise and sell the product on foreign markets under the name "vodka" should belong exclusively to the Poles, the producers of Wódka Wyborowa, Kristall and other brands. Meanwhile, it was asserted that Moscow Special and Stolichnaya, as well as Krepkaya, Russkaya, Limonnaya, Pshenichnaya, Posolskaya, Sibirskaya, Kubanskaya and Yubileynaya, all of them available on the world market, could no longer be termed vodkas, and some other name would need to be found to describe them on their labels and in advertisements.

At first no one in Soyuzplodoimport paid serious attention to this threat, since it seemed completely absurd to suppose that fraternal Poland would make such a demand. The Polish claim seemed no more than a perverse joke. Soviet foreign trade officials were convinced that the entire world knew that vodka had been produced in Russia since time immemorial, and that as a result Russian vodka could not be deprived of its historic, popular, national name through the whim of an unexpectedly stubborn ally.

But the laws of the world capitalist market are ruthless; they take neither emotion nor tradition into account. They demand either formal documentation or other historically convincing proof, establishing this or that date for the invention or first production of an item; this date then confers the prior right of one or another proprietor to the ownership of the given invention or product. These requirements are the same for a great power as for a small country, and they leave no room for either historical tradition or customary practice where these are not backed up by historical argument. For this reason the pleas of the Russian side that the world knew the real situation, and that things had always been that way, were summarily rejected.

The Western European precedents in this regard were completely unambiguous. There were well established dates for the first production of all of the European types of strong spirits: for example, cognac in 1334, English gin and whisky in 1485, Scotch whisky between 1490 and 1494, and German *Bräuwein* (later called *Branntwein*) between 1520 and 1522.

It was therefore considered that there was no case for making an exception for vodka. The date of its initial production had to be established both for the USSR and for Poland. It was entirely possible that, as in the case of English gin, or Scotch and Irish whisky, this would make it possible to establish the prior right of one or the other of the parties to the dispute. That is where things stood at the beginning of 1978, when the contending sides were granted time to conduct research and come up with proof of their claims.

Meanwhile, neither Soyuzplodoimport nor, in particular, most of the leading figures in the Ministry of Foreign Trade, understood the seriousness of the situation or the scale of the task confronting them. People in these institutions supposed that the question of prior right to the name "vodka" would be easily settled; it would be enough to assign a few bibliographers and other specialists to search out the required date in the historical texts or works on spirituous liquors, and there would be an end to the problem. The question thus appeared merely to be one of knowing the techniques and setting to work. However, six months of searching revealed not only the absence of any recorded date for the first production of vodka, but also many gaps in the literature of its history. Information about the invention of vodka could not be found even in the state archives, where there were no authentic documents revealing when distilling first began in Russia. At this point people recognized that the question was exceedingly difficult, that it could not be resolved through the efforts of the Ministry of Foreign Trade alone, and that there was no option but to refer the matter to experts, both in Russian history and in that of the liquor industry.

Consequently, Soyuzplodoimport turned to two leading research institutes, the Institute of History of the Academy of Sciences of the USSR, and the Higher Scientific Research Institute of the Fermentation Products Division of the Central Department of Distilling of the Ministry of the Food Industry of the USSR, with a request that they compile historical references on the point at issue.

Unfortunately, neither a senior researcher at the Institute of History, Doctor of Historical Sciences M. Ya. Volkov, nor the research group of the Ministry of the Food Industry, set up under the director of the Institute and involving G.I. Yushchenko and

Candidate of Technical Sciences V.G. Svirida, could provide any meaningful answers to the question, let alone furnish conclusive proof.

"Instruction 04.13" was the designation given to the secret directive of the Ministry of Foreign Trade "to study the question of the priority of Russia in the production of vodka". Both research groups, that of the Academy of Sciences and that of the Ministry of the Food Industry, came up with stock answers that proved only one thing: for many years they had been consuming state resources to absolutely no purpose.

After both of these official state institutions with their hundreds of staff had proved unable to provide the government with even elementary assistance, the management of Soyuzplodoimport turned on its own initiative to the present author who, despite not being on the staff of any state institution, nevertheless agreed "as a matter of civic duty" to undertake objective historical research to establish when it was that vodka first appeared in Russia as a completely new, previously unknown product.

Thus this work came to be written essentially as a research monograph. Originally its sole purpose was to establish the date when vodka was first produced. Since the traditional kind of documentation from administrative and fiscal sources was lacking, the study was conducted from an angle which is unusual for present-day historical works – that is, through an analysis of the terminology of distilling, through research into the significance of the term "vodka", and through examining the historical conditions in which vodka might have appeared.

These characteristics of the research naturally determined not only the plan of study but also the course of the research, and even to a significant degree the form of the analysis and the methods used by the author in proving his assertions. Less patient readers may well find the argumentation slow-paced, cavilling and pedantic. On points where everything might already have seemed clear, the author nevertheless asks himself new questions, advances new arguments or objections against already proven conclusions, in order, through overturning these new objections, to show that it was not possible for any other answer to be arrived at. Such methods, of course, involve digressions; but they helped to answer the questions and doubts of the arbiters who were required to familiarize themselves with the Polish and Soviet submissions on the production of vodka in their countries. Furthermore, in order to avoid the need during the arbitration process for any hasty additions and clarifications, the author decided to include in his basic research all the conceivable and even inconceivable objections which his work might have encountered.

The limited work performed by historians on the question at issue, the uncertainty as to which century had seen the first production of vodka, and the difficulty of deciding where to begin the search for the origins of vodka forced the author of the present text to undertake a plan of work which was difficult, but the only logical and soundly based one in the circumstances. This was to analyse scrupulously, step by step and in chronological order, the entire history of the development and production of spirituous liquors from the moment of their appearance in our country, and in this way sooner or later to happen upon the date of the first production of vodka. This method, like others that were employed, did not conduce to brevity and clarity, or to an exclusive concentration on answering the question of when vodka was first distilled. Other material had to be examined in passing – for example, texts on such alcoholic drinks as ale, beer and so on. The procedure gave the work a somewhat prolix character, and suggested that the author had been distracted from the basic question of vodka. But for the matter in hand, for the purpose of revealing the truth, this method turned out to be extremely useful and productive, since it made it possible to define the differences between vodka and brewed alcoholic drinks – *kvas*, ale, light beer and beer – and thus made it easier to define the date of the first production of vodka.

In short, the materials in the present work which do not bear directly on vodka are not superfluous or irrelevant to the book's purpose of revealing the history of vodka.

The author's research not only answered the question of when vodka first appeared, but also suggested why this occurred at a particular time rather than sooner or later. The research also provided answers to questions surrounding the first production of strong alcoholic liquors in the Ukraine, Poland, in Sweden and

Germany. The dates arrived at for these countries were in line with those submitted to international tribunals by foreign historians. The researchers acting on behalf of the Polish People's Republic did not succeed in showing that *gorzalsia* (the original Polish name for vodka) was produced in the Cherkassy region before the middle of the sixteenth century. In fact the present author's data shifted the date of birth of vodka in Poland back some ten to fifteen years before the date arrived at by the Poles themselves; but this was still significantly later than the first date for the production of vodka in Russia.

In 1982 the international tribunal decided in favour of the Soviet Union, establishing beyond question our status as the inventors of vodka and securing for us the right to advertise it on the world market under this name. Thus an attempt by certain foreign circles to harm Soviet commercial interests and to use the question of vodka for scandalous and disreputable ends, including that of prejudicing Soviet–Polish relations, ended in failure.

Readers should bear this in mind as they peruse the materials that follow. Of course the Soviet Union was formally wound up at the end of 1991 and its assets made over to a Commonwealth of Independent States, including the Republics of Russia, the Ukraine and Belorussia. I do not think that these developments lessen the intrinsic interest of the research reported here. It is to be hoped that the Commonwealth and its constituent parts will make a wise use of their inheritance from the old Union in this as in other domains. Indeed it is a central theme of the present study that vodka has always furnished vital tests to those who would like to rule Russia. In Chapter 5 the reader will find reflections which go beyond the report prepared for the Ministry, and which develop the hypothesis that the passage of control of vodka from the hands of one social group to that of another has marked the successive stages of our history. I would venture to suggest that this may prove to be the case in the future as it was in the past.

W.V. Pokhlebkin

1

The Origins of Alcoholic Liquors in Russia

Vodka in Its Social Context

The history of any product forms a part of how a society repro-
duces itself and of how its material presuppositions develop.
Through studying the origins of particular products we can
discover how these material conditions came about; how the
products relate to the prevailing pattern of production and repro-
duction; and the reasons for that relation. This allows us to
arrive at a better understanding of the social relationships which
govern a society.

The history of a product thus appears as one of the atoms
which make up history as a whole, one of the "building blocks"
out of which the edifice of human history is constructed. It is
clear that certain products have played important roles in the
history of humankind, either at particular stages of human development, as for example with spices, tea, iron, petroleum and
uranium; or throughout the whole recorded history of humanity,
as with bread, gold and alcoholic liquors.

Among the multitude of products which humanity has created
or used, vodka – or, to use a more general term, grain spirit –
occupies a special position due to its diverse impact on human
society, on people's relationships with one another, and on social
problems. The three main problem areas that emerged with the
advent of spirituous liquors – the fiscal, the productive and the
social[1] – have not only persisted over the centuries, but show a

tendency to grow and become more intractable as human society develops and becomes more complex.

Consequently, the history of vodka is not by any means a trivial or lowly topic; it is far from being a minor footnote to human history. The issue of vodka is important enough for serious historical and scientific work in this area to be an urgent necessity, and vodka has in fact already become the topic of an extensive literature. Unfortunately, this literature is devoted almost exclusively not to the history of the product itself, but to the consequences of its use; though admittedly these consequences of vodka are particularly striking.

Through examining the consequences of a phenomenon one can never come to understand its whole scope or its essential nature. From this flows the extremely contradictory character of everything that has been written about spirituous liquors, beginning with their physiological effects and ending with their social and historical significance. What we generally find here is an extreme polarization of views, of the "useful vs. harmful" variety. In the case of vodka, there is much confusion as to the circumstances of the emergence of the drink, and consequently about the development of vodka and vodka production as processes closely linked to the general course of historical evolution. As a result, there is a failure to bring out the diverse significance of vodka in various societies at particular stages of their history. In sum, as soon as the discussion turns to vodka, Marx's insistence that the truth is both concrete and historical is forgotten.

In much previous writing on spirituous liquors, and especially on vodka, the cause of objective historical inquiry has been done a grave disservice by a complete disdain for such objective considerations as terminology and chronology. In various sources the term "vodka" is applied to completely different products. The term is also used in the context of historical epochs in which there was simply no such thing as vodka. This is why the first step in a study of the history of vodka must be to establish the terminology and chronology in precise fashion. Only after this can one address the root of the question: where, when, and under what circumstances did vodka emerge? Why were the conditions for the development of production most propitious in one particular location rather than another, despite the fact that in later times the development of technology would make the production of vodka possible at practically any point on the globe?

An examination of the terminology and chronology must proceed separately from a study of the "recipe" or methods of production of vodka, since different ingredients, formulas and techniques may at times be described in the same terms. It is for this reason that a determination of the "date of birth" of vodka requires a combination of all three "dimensions" – the terminological, the chronological and the technological – and their comparison with one another. Only on this basis can precise and convincing results be obtained. It is therefore indispensable to establish in advance what our range of sources is to be, and to identify the criteria we shall use for assessing the worth of the information they provide. This range may be summarized as follows:

- Archaeological materials.
- Written documents of the period from the fifteenth to the nineteenth centuries, including chronicles.
- Evidence from folklore: songs, proverbs, sayings, tales and fables.
- Cookery books of the fifteenth to the eighteenth centuries.
- Dictionaries of various types, including etymological, explanatory, technical and historical, and lists of foreign words.
- Literary evidence: historical research, literature, diaries and memoirs of contemporaries.
- Specialist literature, such as technical and pharmaceutical writings.

Of this list the most valuable sources are documents and dictionaries, especially etymological and linguistic-historical dictionaries summarizing the linguistic materials in texts from the eleventh to the seventeenth century.

The least reliable sources are chronicles and folklore. In the first case this is because the chronicles were written as much as three or four hundred years after the events described, and repeatedly revised. In the process of rewriting, everything that did not relate to the chronology of political history was much changed. This was especially true of lexical material: the terminology was modernized, and at times the old terms acquired a new content. A comparable process went ahead even more powerfully in folklore, on which it is almost impossible to rely as a trustworthy source. The recording of folklore took place mainly at the end of the nineteenth and the beginning of the twentieth century, and it was occasionally carried out in an insufficiently critical fashion. Besides, attention was concentrated on the narrative rather than the lexical details.

Among the other sources, archaeological materials such as relics of daily life are extremely scarce and of only indirect significance. These might have provided the most compelling evidence of all. But unfortunately, excavations have so far failed to turn up a vessel containing spirituous liquor, or a contemporary still, such as might have given us precise knowledge of the nature of the product and the manner of its production.

Written sources, and especially the works of historians, contain extensive and mainly reliable material. But none of these sources is devoted specifically to the history of spirituous liquors, let alone vodka in particular. Such general works as *The History of the Russian Tavern* or *The History of Drunkenness* concentrate their attention basically on social issues and are hardly at all concerned with the names and characters of the various drinks. Despite the abundance of literary and other sources there is only a meagre amount of material that is both dependable and useful, which renders research extremely difficult.

It follows that the only scientific approach to solving the question of the origins of vodka as Russia's national spirit is through the painstaking comparison of all available materials along the three lines I have mentioned.

The Terminology of Alcoholic Liquors

Before we begin to examine the terminology of alcoholic beverages that existed in our country from the ninth to the twentieth centuries, it should be stressed that scholarly precision and honesty demand that we should not allow our modern point of view and our contemporary understanding of the meaning of words to intrude into our analysis of the old language. We should not allow ourselves to fall into errors concerning words which by chance have the same sound and even the same spelling as modern words, but which in the remote past or even two or three centuries back had meanings quite different from those they possess today. Therefore, in order to avoid confusion in our search, and to understand clearly the meaning and connotations of a particular name, we must examine the senses in which each term was used at different times. For this reason one and the same word or name may be explained several times over, separately for each period and, of course, with several different meanings.

The Meaning of the Word Vodka

Here we have to examine not only when the word first appeared in Russian, but also whether it existed in other early Slavonic languages.

The word vodka, in its contemporary sense of a strong alcoholic drink, is widely known both in our country and abroad; yet few people know its real meaning. In fact it means nothing other than water (in Russian, *voda*), but in the diminutive form. Diminutives now tend to have a rather coarse ring to them, but they are part of the authentic popular language. However, we no longer perceive this sense of the word, at once endearing and vulgar; we no longer hear a diminutive of *voda*, but a completely independent word: vodka.

Another word relating to food and drink, and almost as ancient as *voda*, is *ovoshch*, vegetable. This lost its diminutive form around the beginning of the eighteenth century; dictionaries from the end of the century were forced to explain that *ovoshchka* meant a small vegetable.[2] As is clear from comparing the words *mama/mam-ka*, *ovoshch/ovoshch-ka*, *voda/vod-ka*, the diminutive may be formed by the direct addition of the suffix *-ka* to the word root. Other diminutive forms of the word *voda* have survived – *vodichka*, *vodon'ka* and *vodochka* – of which the last mentioned has also come to be associated with alcoholic liquor rather than water.

The fact that the name vodka has its origins in the word *voda*, and hence is somehow connected with water, is thus proven beyond doubt. This is extremely important for establishing the nature of vodka as an alcoholic drink, as we shall discover later in the context of the history and technology of vodka production.

The reader will not be surprised to learn that none of the etymological dictionaries of the Russian language discuss the word vodka,[3] since etymological dictionaries are concerned only with the origins of the primary, independent root forms of the words which make up the language, and do not deal with derivative forms. The linguistic specialists who compiled these works did not examine the word in its present-day sense, and merely noted it as a modification of *voda*. Hence they cannot help us to determine when the word vodka acquired its present meaning, nor can they tell us what influenced this process.

However, there are also explanatory dictionaries of the Russian language which explain the meanings which are current for words at the time of compilation. What do these works have to say about vodka? The largest, most exhaustive, and in general the best of these explanatory dictionaries is the nineteenth-century one of V.I. Dal. But, despite its scope, this work does not list *vodka* as an independent lexical item. The contemporary meaning of the word is examined under the heading of the word *vino* (wine), while the dictionary assigns *vodka* the same meaning as *voda*.[4] From the date of the dictionary we can deduce that the word vodka was not widely distributed before the second half of the nineteenth century, although it was already known and used among the common people. It is only in dictionaries published toward the end of the nineteenth century and in the early years of the twentieth that the word is regularly encountered as an independent lexical item, with its sole contemporary meaning of a strong alcoholic liquor.

Meanwhile in dictionaries of regional usage, dealing not so much with the general Russian lexical stock as with provincial dialects, *vodka* is listed as having only one meaning, its original one of water. The connotation of an alcoholic liquor is totally absent even from the most thorough of these works, Filin's dictionary.[5] This exhaustive record of all the dialects of Russian reflects the usage of the middle of the nineteenth century (the material was compiled between 1846 and 1853) and deals with the language of the population in a broad territory to the east and north of Moscow – the Vladimir, Kostroma, Yaroslavl, Vyatsk and Arkhangelsk provinces.

This shows that in the middle years of the nineteenth century the word vodka, in the sense of an alcoholic liquor, was known in Moscow, in Moscow province, and in the provinces of the grain belt where distilling was originally developed: that is, in Kursk, Orlov and Tambov provinces, and in the so-called "free Ukraine" around Kharkov and Sumy.

It becomes clear that the middle of the nineteenth century constituted a critical juncture, when the word vodka with its present meaning had begun to acquire a broad currency in the general Russian language, but when this process had still not been completed. We thus have an indication that the origins of the new meaning of the word to signify a spirituous liquor should be sought at some earlier time. The middle of the nineteenth century marks the latest point of its introduction, at which it had already become established. But the earliest point has yet to become clear.

We must now turn not to dictionaries of the contemporary language, but to the dictionary of Old Slavonic; that is, to the language of the numerous chronicles of the period from the ninth to the thirteenth centuries.[6] This dictionary, compiled on the basis of a painstaking examination of all the texts in Old Slavonic which survive in Czechoslovakia, Bulgaria, Serbia, Poland, Lithuania, Belorussia, Moldavia and the Ukraine, does not contain the word *vodka* in any form. It is quite impossible that this dictionary, prepared over several decades by the world's greatest Slavists, could have omitted a particular word, since the word list and card index on which it was based contained all the words of all the texts of the ninth to the thirteenth centuries; and also from a significant section of the texts of the fourteenth century, which saw the end of the gradual transition from the Old Slavonic language of all the Slavs to the Slavic national languages. This indicates that, at least up to the beginning of the fourteenth century, the word vodka, both with the meaning of water and with that of alcoholic liquor, was absolutely unknown both in Russia and throughout the Slavic world.

The foregoing allows us to draw the following tentative conclusions.

The word vodka with the meaning of alcoholic liquor appeared in the Russian language no earlier than the fourteenth and no later than the middle of the nineteenth century.

The word vodka with the meaning of water – that is, the diminutive form of *voda* – did not exist in the common Slavonic language at least until the thirteenth century and perhaps until the fourteenth. If follows that the diminutive meaning arose wholly within Russian at the stage when this was beginning to form itself into a national language; that is, during the thirteenth and fourteenth centuries. The Ukrainian language began to appear as a national tongue during the fifteenth century, and does not include the word except as a much later importation. As a result of the Tatar invasions, Russian developed in isolation after the first three decades of the thirteenth century, shut off from the influence of Western languages – notably that of Latin, which had a significant effect on Polish. The Russian language therefore had its own, quite distinct forms of development.

It is thus clear that the word vodka, in any sense and irrespective of the time of its appearance, is a native Russian word. Its appearance in other Slavonic languages can be explained only as the result of later borrowings from Russian, borrowings which did not occur earlier than the fifteenth or early sixteenth centuries.

The fact that the word vodka with the meaning of alcoholic liquor was absent from the Russian language before the fourteenth century does not prove that at this time there were in Russia no spirituous liquors of different names and produced in different ways, or even strong alcoholic liquors similar to vodka in the technology of their production but with different names both for themselves and for their production techniques and equipment.

In the same way, one cannot directly link the presence in the language of a term with the presence of a product corresponding to the modern meaning of that term. Such a product might have existed under a different name, or it might not have existed at all. In this area as well, proof is required. First of all we must establish which terms denoting alcoholic liquors were employed in Old Russia, and what these terms signified.

Old Russian Terms for Alcoholic Liquors

In the period between the ninth and fourteenth centuries a number of terms were used in Russia to denote drinks. These were:

pivo (drink), *voda* (water), *syta* (honey water), *berezovitsa* (birch-sap wine), *vino* (wine or spirits), *med* (mead); *kvas* (a drink having much in common with beer); *sikera* (beer in general); and *ol* (ale). Most of these, other than *voda* and *syta*, were alcoholic and intoxicating; *berezovitsa* had two forms, plain *berezovitsa* and "drunken" *berezovitsa pianaia*. It was the same with *kvas*. The borderline between alcoholic and nonalcoholic drinks was thus extremely vague. Even *syta*, a mixture of water and honey, could easily ferment and become transformed into a weakly alcoholic drink, while retaining the name of the nonalcoholic variety. Grape wine (*vino*), which was imported from Byzantium and the Crimea, was mixed with water in the same fashion (as was the ancient Greek custom). It is easy to understand how the word *voda* came to be closely associated with alcoholic liquors, and why it had the sense of a beverage rather than merely of water in a general sense, as we think of it today. When we come to discuss the reasons why vodka, one of the strongest alcoholic liquors known to the Russian people, was given the name of such an innocuous drink, we need to bear in mind how different a view of water medieval Russians had from that of our contemporaries; for they saw it as the basis of many or even all drinks, including of course all alcoholic ones.

We must also keep in mind the fact that in the period from the ninth to the eleventh centuries the water which was recognized as a drink was not any water, but only "running water" (*voda zhivaia*);[7] that is, water from springs and clear, fast-flowing streams. From the twelfth century this term was increasingly replaced by "spring water" (*kliuchevaia voda* or *rodnikovaia voda*), and by the middle of the thirteenth century it had disappeared from the everyday spoken language, persisting only in folk tales. There the term lost its real meaning as this was forgotten by the people, who reinterpreted the phrase in a symbolic spirit, opposing living water to dead. There is no doubt that at the time when vodka first appeared people still recognized the old meaning of the term "living water", though they did not use this in daily life. Therefore the new spirituous liquor was not given the name "water of life" or "living water", as happened universally in Western Europe and among the western Slavs, who were subject to Latin influences. Throughout this region the earliest distilled spirit, containing half or less by volume of water, was given the name *aqua vitae* (water of life), from which are derived the French *eau de vie*, the Scots *whisky* (via the Gaelic *uisge beatha*), and the Polish *okowita*. These names are simple translations from Latin; or else, as in the Scandinavian *akvavit*, the Latin phrase has been taken directly into the national language.

This did not occur with the Russian language, since the production of vodka did not have Latin or Western European origins but quite different ones, partly Byzantine and partly native Russian. This is why the name *aqua vitae* was not reflected in the Russian terminology of spirituous liquors, either before or after the thirteenth century. Meanwhile, the term "living water" in Russian referred only to plain drinking water.

"Living water" in Russia was also referred to as *pivnaia voda* (water for drinking),[8] and sometimes simply as *pivo* (drink). These terms too helped to strengthen the association between *voda* and other drinks, including *pivo* in its present sense of beer. Through this process, water was placed in the same category as other beverages. But simultaneously, water was understood symbolically as the diametric opposite of alcoholic liquors.[9] For this reason it did not occur to anyone to give a strong alcoholic drink the name "living water"; that is, to draw a parallel between alcohol and running water. Vodka received its original name not by analogy with water, but with wine, the most ancient of intoxicating liquors.

It is evident that genuine vodka appeared on the scene before its present name emerged. This can be seen from an analysis of the terminology associated with various drinks. Significantly, vodka was referred to as *vino* (wine) for a very long time, right up until the beginning of the twentieth century, when its present name at last became firmly established. It is thus logical to suppose that vodka existed under the name *vino*, or perhaps under some other name, long before the current name vodka became attached to it.

If we are to resolve this question, therefore, it is extremely important to analyse in detail all the terms used to refer to alcoholic liquors in Old Russia, in the Novgorod, Kiev and, to some extent, in the Vladimir-Suzdal regions, before the advent of the name vodka.

Vino: from the ninth to the thirteenth centuries this term, where it was used without adjectives, referred only to grape wine.[10] It was known in Old Russia from the ninth century; after the advent of Christianity at the end of the ninth century, grape wine became obligatory in the ritual of the Eucharist.[11] It was imported from Byzantium and Asia Minor, and was known as "Greek" or "Syrian". Until the middle of the twelfth century it was drunk only in a mixture with water, as it had traditionally been drunk in Greece and Byzantium.[12] One source enjoins its readers to "mix water into wine",[13] rather than adding wine to a container of water. The name *vino*, from the Latin *vinum*, was adopted when the Gospels were translated into Old Slavonic.

By the middle of the twelfth century *vino* had already acquired the meaning of pure grape wine, without any admixture of water. To avoid confusion, it became obligatory in both the old and new terminology to note all cases in which the wine referred to was not pure: "And the *arkhitriklin* [the organizer of the banquet] did taste the wine, which was of water."[14]

In order to avoid long explanations, writers very often used an adjective to specify which type of wine they were discussing. Thus, terms appeared such as *ots't'no vino* (vinegar wine), for a sour, dry wine;[15] *vino osmr'neno* (tempered wine), for a sweet grape wine with spices;[16] and *vino tserkovnoe* ("church wine"), for red grape wine of a sweet or dessert type, of good quality and without added water. Finally, in a document dating from 1273, the term *vino tvorenoe*, "prepared wine", first appears in the written sources.[17]

A detailed examination of this term will be made later; here we should note that it arose almost 400 years after the advent of grape wine, and some 200 to 250 years after various adjectives describing distinct types of grape wine had come into use. This in itself indicates that what we are dealing with here is not grape wine, not wine obtained through a natural process, but a beverage produced in some different and artificial manner. The term no longer refers to wine as it was understood before the thirteenth century. It was a different drink, made from different materials, as we shall see.

Med (mead): This is considered to have been the second most important alcoholic beverage of Old Russia. Honey had been familiar from ancient times both as a sweetener (Latin *mel*) and as the key ingredient in an alcoholic drink (Latin *mulsum*). Mead was drunk by most of the peoples of the European forest zone, including the ancient Germans, who called it *meth*; the Scandinavians, who called it *mjöd* and regarded it as the drink of the gods; and especially the ancient Lithuanians, who knew it as *medus*. The origins of the word are not Russian, but Indo-European. In the Greek language the word *methu* denoted any intoxicating drink; but it was sometimes used with the meaning of pure wine, that is, wine which was too strong to be drinkable according to Greek ideas and traditions. The word *methus* meant drunken. All this testifies to the fact that the alcoholic content of mead was much greater than that of grape wine, so that both the Greeks and the Byzantines considered the use of such strong liquors to be characteristic of barbarians.

Meanwhile in Old Russia, folklore indicates that mead was the most widespread alcoholic drink; wine is hardly mentioned. Documentary sources, however, suggest the reverse. They record the use of imported wine; but the first mention of honey, used as a sweetener, dates from the year 1008 in Old Russia,[18] and from 902 in Macedonia. The first mention of mead as an alcoholic drink dates from the eleventh century in Lithuania and Polotsk,[19] in Bulgaria from the twelfth century, in Kievan Rus from the thirteenth century (1233),[20] and among the Czechs and Poles from the sixteenth century. An earlier mention occurs in the chronicle of Nestor, which records that in 996 Vladimir the Great ordered the preparation of "300 vats of mead for a feast following the victory over the Pechenegs".[21] At the beginning of the tenth century the Arab traveller Ibn-Dast (Ibn-Rustam) had also noted that the Russians had an intoxicating drink made from honey.[22]

From the above we can draw the following conclusions: mead, as an alcoholic liquor, was first made and was most familiar to the population in the wooded areas of Old Russia, on the territory of present-day Belorussia, and in the principality of Polotsk, areas in which honey was gathered from the hives of wild bees in the forest. From these regions mead was later taken into Kievan Rus along the Pripyat and Dnepr rivers.

In Kiev in the tenth and eleventh centuries mead was drunk only on special occasions, for which it was brewed from stocks of honey by a process involving heating. This brewed mead was of a lower quality than the so-called "matured mead" (*med stavlen-*

nyi). The latter was the product of a natural, cold fermentation of honey together with the juice of berries such as bilberries and raspberries, and was aged for ten or fifteen years, sometimes longer. At princely feasts in the fourteenth century mead as much as thirty-five years old was sometimes served. Since the general spread of both types of mead took place in the period between the thirteenth and fifteenth centuries, the notion that in ancient times the principal drink was mead was reflected accurately in folklore; although this was created in a somewhat later period when Russian national culture had begun to take shape.

The expansion of mead-brewing between the thirteenth and fifteenth centuries came not because it was a new drink — it had reached most regions between the ninth and the eleventh centuries — but with the decline of imports of Greek wine after the Mongol-Tatar invasion and of the collapse of the Byzantine empire. Hence historical circumstances, including not only changes in international relations and international trade, but also geographical adjustments — the shifting of the territory of the Russian state to the north-east, and the relocation of the capital from Kiev to Vladimir and then to Moscow — led to changes in the nature of the alcoholic beverages that were consumed in these places. Russia became remote from the sources of grape wine, and its people were forced to seek local ingredients and devise local methods of producing alcoholic liquors.

Between the thirteenth and the fifteenth centuries this ancient drink acquired new and considerable importance in the everyday life of the well born and wealthy. The long period needed for the production of genuine, high-quality matured mead undoubtedly priced it out of the reach of most would-be consumers. Even at the court of a great prince, the commoners would have to make do with a cheaper, more rapidly prepared, but more intoxicating drink: brewed mead.

The thirteenth century thus marked the transition first to drinks prepared from local ingredients, and second to drinks that were significantly stronger than those which had been consumed during the preceding five centuries. There can be no doubt that the habit of consuming more intoxicating liquors also prepared the way for the appearance of vodka.

A developed and extensive mead-brewing industry inevitably called for the use of distilled spirit as a component of cheap but strong mead. Already in the fifteenth century honey was becoming much less freely available; it was in keen demand in Western Europe, and as it became an object of export its price rose.[23] It was necessary to find a cheaper and more widely available raw material for local use. This raw material was grain, which from the earliest times had been used to produce drinks such as *kvas*.

Kvas: this word is encountered in Old Russian texts beside references to wine, and earlier than any mention of mead. Its meaning, however, does not correspond fully to that of the same word in modern Russian. In a text from the year 1036 we find a clear mention of *kvas* as an alcoholic drink, since in the language of that time the word *kvasnik* had the meaning of drunkard.[24]

In the eleventh century *kvas* was brewed like mead, and therefore resembled *pivo* (beer) in the present sense of that word; but it was thicker and more intoxicating. Later, in the twelfth century, a distinction began to be made between *kvas* as a sour, weakly alcoholic drink and *kvas* as a powerfully intoxicating one. Both, however, kept the same name, and it is only possible to guess from the context which type is being referred to.

In the second half of the twelfth century or near its end strongly alcoholic *kvas* began to be called *tvorenyi kvas*; that is, "prepared" *kvas*, specially brewed, and not simply left to ferment like ordinary *kvas*.[25]

This brewed *kvas* was considered to be an alcoholic beverage as strong as pure wine. "You should not drink wine or brewed *kvas*", states a church edict.[26] "Woe to those who brew *kvas*",[27] we read in another source. It is plain that what was involved here was no inoffensive drink. Of all the varieties of brewed kvas the strongest was *kvas neispolnennyi* – "unfinished" *kvas*, which was often also described as "lethal".[28] In the Old Slavonic language the word *neispl"nenyi* meant uncompleted, not ready, of poor quality – the opposite of the Latin *perfectum*.[29] This indicates an inadequately or badly brewed product, which must have contained a large amount of the highly poisonous heavy alcohols known as congeners. The word *kisera*, which occurs only rarely in the sources and which denotes a powerfully stupefying drink, also apparently refers to this type of *kvas*. If we reflect that the word *kvas* is cognate with *kisloe* (sour), and that the drink was sometimes referred to as *kvasina*, *kislina* and *kisel*, then *kisera* can be seen as a derogatory form denoting *kvas* which was unfinished, spoilt, or bad.

There are indications, however, that *kisera* may have been a corruption of the word *sikera*, which also referred to one of the old alcoholic liquors.

Sikera: this word existed in the Russian language until the fourteenth or fifteenth century, the period that saw the change in the methods of production of alcoholic beverages in Russia and in the associated terminology. Then the word vanished without trace from both spoken and written language, leaving no substitute or analogue. We shall attempt to elucidate its meaning and original sense, since this will cast light on the history of Russian alcoholic liquors.

The word *sikera* entered the Old Russian language from the Gospels, where it was reproduced without translation from the Greek since translators at the end of the ninth century could find no equivalent for it in the Slavonic languages, including Old Russian. The word was used to refer to alcoholic liquors in general, but at the same time was clearly differentiated from grape wine. "Wine and *sikera* should not be drunk."[30] In the Greek language of the Gospels the word was used for brewed intoxicating liquors in general, including any alcoholic beverage apart from wine.[31] However, the origins of this word lay in the Old Hebrew and Aramaic languages – *shekar* (or *shekhar*) and *shikra*.

Shikra in Aramaic signified a kind of beer, and it was this word which entered the New Testament as *sikera*.[32] *Shekar* in biblical Hebrew meant any alcoholic drink other than grape wine.[33] This word passed into Russian as *siker*. Therefore in some sources we encounter *sikera*, and in others *siker*. The fact that these words were so alike in sound and meaning led even linguists to treat them as variants of the same word. But in fact they were not only different words but had different meanings, if we consider them from a technological point of view.

In Palestine and among the Greeks *sikera* was a strong liquor made from the fruit of the date palm. In its Aramaic sense *sikera* signified an intoxicating alcoholic liquor the technology of whose production was close to that used to prepare mead or beer, without distillation.

There can be no doubt that in Russian monasteries learned monks discovered the true meanings of the Hebrew and Aramaic words mentioned in the Old and New Testaments, and thus gained insight into the products and their manufacturing processes.

Pivo: as well as the alcoholic beverages listed earlier – wine, mead, *kvas* and *sikera* – sources from the ninth to the eleventh centuries very often mention *pivo* (beer). However, it is clear from texts of the time that *pivo* originally referred to drinks in general, and not to an alcoholic drink of a particular kind as it does today. "Bless our food and *pivo*", we read in an eleventh-century chronicle.[34] Later, however, we encounter the term *tvorenoe pivo*, that is, a specially brewed drink. Brewed *pivo*, it is clear from the sources, was very often referred to as *sikera*, and sometimes went under the name of another drink, *ol*. The term *pivo* continued to be used with a broad meaning throughout the twelfth and thirteenth centuries. But while in the tenth and eleventh centuries the name had been applied to any drink at all, in the twelfth and thirteenth centuries people began using it to refer to alcoholic drinks produced artificially by human agents.

Ol: In the middle of the thirteenth century there appeared for the first time another term denoting yet another alcoholic drink – *ol* or *olus* (ale).[35] There is also evidence that in the twelfth century the term *oluy*[36] was used; this appears to have had the same meaning as *ol*. To judge from the meagre descriptions contained in the sources, *ol* referred to a drink similar to the contemporary *pivo*. This beer or ale was prepared not simply from barley, but with herbal admixtures such as hops and wormwood. *Ol* was therefore sometimes called *zel'e* or *zele*, from the root *zel-* (green). Its name recalls the English *ell*, which was also prepared from barley and herbs, including, for example, heather flowers. The fact that *ol* later came to be identified with *korchazhnoe pivo*, "pot beer", indicates strongly that the drink known in the twelfth and thirteenth centuries as *ol* was similar to the *pivo* of those times.

It is evident that the term *ol* was used to denote a relatively strong drink of high quality, probably resembling porter. At the end of the thirteenth century it was stated in the *Nomokanon* that *ol* could be brought into a church "in place of wine"; that is, could be used in the Eucharist. None of the other drinks of the period was accorded this distinction.[37]

Berezovitsa pianaia: This term is absent from the Old Slavonic texts, but from the reports of the Arab traveller Ibn-Fadlan, who visited Russia in the year 921, we know that the Slavs drank *berezovitsa pianaia,* made from the naturally fermented sap of the birch tree. This was stored for long periods in open vats, where it turned into an intoxicating liquor.[38]

An analysis of the terminology associated with alcoholic liquors between the ninth and fourteenth centuries allows us to be sure that in the early historical epoch five types of alcoholic beverages were consumed in Russia.

Grape wine, mainly red, was obtained from Byzantium and the countries of the Mediterranean. Until the thirteenth century all types of wine were included under the general term *vino,* sometimes with the adjectives *ots't'no* (sour, vinegar) and *osm"r"neno* (sweet, dessert, spiced).

Drinks were obtained through the natural fermentation of local products – birch (or sometimes maple) sap, honey and the juice of berries – without any additional human action; that is, without the addition of yeast, and without brewing. These drinks were *berezovitsa pianaia* and *med stavlennyi.*

Drinks were made by the artificial fermentation of grain products – rye, barley and oats – after the boiling of the wort and with the addition of herbs such as hops, St John's wort and wormwood to enhance the flavour. These drinks were *kvas* – the common beer of the period – and *ol,* a strong, thick beer.

Drinks were obtained through the artificial fermentation of honey, or through combining artificially fermented mead with the products of artificially fermented grain. One such drink was *med varenyi* (brewed mead), or simply *med.* It was made by dissolving honey in water; this solution was then enriched with malt, flavoured with various herbs – again hops, wormwood and St John's wort – and brewed in a similar way to beer. The resulting drink had a relatively high alcoholic content, since the honey wort was rich in sugar; "brewed mead" was more intoxicating than *ol.*

Drinks were made by distilling fermented grain products. In this category were *kvas tvorenyi, vino tvorenoe, sikera* and *kvas neispolnennyi*. It appears that all these words referred to one drink. This was given various names in different sources, first of all because in the period from the eleventh to the thirteenth centuries it was a quite new beverage, which had appeared after all those listed earlier; a term for it could be chosen only by analogy with the old names designating already known alcoholic drinks. Also, the ingredients of this new drink were the same as those of earlier drinks, however different the outcome was; people were still accustomed to describing the product in terms of the raw material, and not according to the results of processing. The fact that the names referred to one and the same drink is indicated by the single adjective used to describe it – *tvorenyi* (prepared), which points to the uniform nature of the technology employed. It is evident that here is the first grain spirit obtained by the distillation of the fermentation products of ingredients rich in sugar and starch.

Thus our review of the terminology assures us of the existence of several distinct processes, including the natural fermentation of mead; the brewing of mead; the brewing of *kvas* and beer; and the "preparation of wine" (*vinotvorenie*), that is, something close to *vinokurenie*, distilling.

Russian Terms for Alcoholic Liquors in the Fourteenth and Fifteenth Centuries

Earlier we examined the terms designating alcoholic beverages mainly in sources up to and including the thirteenth century, though the vocabulary of the main texts of the fourteenth and fifteenth centuries was also taken into account. We must, however, bear in mind that we know little about the everyday spoken language of this time. Ecclesiastical language was recorded in writing; but this was constantly being cleansed of vulgarisms, and therefore does not fully reflect the vocabulary that was actually used in the period.

It is especially important to be conscious of this because, in historical terms, this period constituted a turning-point in state and political affairs and also in the sphere of production and the economy in general. Naturally, daily life and language were affected. The new epoch, characterized by changed relations with the outside world (resulting from the founding of the Moscow state, the casting off of the Tatar yoke, the fall of Byzantium and the establishment of relations with Western Europe) and by new forms of production (that is, new trades and new products) could not fail to call forth new linguistic phenomena and give rise to new usages.

Although totally new terms are not encountered in this period, one notices changes in the frequency with which earlier terms are employed. The term *khmel'noe*, meaning both "made with hops" and "intoxicating", is used more widely and frequently to denote alcoholic liquors.

For example, the oldest surviving document in the Russian language, the treaty drawn up in 1265 between the city of Novgorod and the Great Prince Yaroslav Yaroslavovich of Tver, speaks of a customs duty being levied on every *khmel'na koroba*,[39] that is, on every unit by volume of alcoholic liquors. A *korob* was a large basket or hamper, perhaps of bark, of such size that it corresponded to a wagonload, since only one could be carried on a cart. The same text states that a duty of two *vekshi* is to be levied on each "boat, wagon and *khmel'na koroba*".

What kind of intoxicating liquor does one measure by the cart-load? Obviously not "matured mead"; it must have been the brewed type. This is indicated by the 1270 treaty between Novgorod and Tver, under which the Prince of Tver was obliged to send to Ladoga an official *medovar* (mead-brewer), and also by the fact that in this period the general term for strong liquor was more and more often becoming *med* (mead), which obviously could not have referred to the matured product. The latter in such contexts was referred to as "Tsar's mead" or "boyar's mead". The term *med* is encountered still more often in texts of the fourteenth and fifteenth centuries, and in fifteenth-century sources is understood to mean a very strong drink intended for the mass of consumers, including the army. This *med* acquired the synonyms *khmel'noy* and *khmel'* (hopped, strong), since it was flavoured with a liberal admixture of hops. The drink possessed "strength" as a result of the congeners which contaminated it, but the term "hopped" was also affixed to it because the more of these congeners were present, the more hops had to be added to mask their unpleasant flavour.

In this period another term came increasingly to be added to the qualifier *khmel'noy*. This term, *zel'e*, referred to the addition of herbs, since, along with hops, wormwood was included among the ingredients. Meanwhile the term *zel'e*, or *krepkoe zel'e* (strong herbs), was steadily acquiring a new sense: it came to denote a viciously intoxicating drink, regarded with disdain for its low quality. Drunkenness itself was clearly beginning to acquire different associations; from being linked with merriment, it began to suggest ruination and the loss of reason.

Thus, for example, the chronicler could not ignore how disastrous the consequences of the "new drunkenness" had been for society in general, for people's behaviour, and for important events. He depicts the surrender of Moscow to Khan Tokhtamysh as due in significant degree to drunkenness in the besieged city on the night of 23 August 1382; a drunkenness that was accompanied by astonishing, incomprehensible recklessness. During the siege "some prayed, while others brought the boyars' mead out of the cellars and began to drink it. Drunkenness emboldened them, and they climbed up on to the walls" to taunt the Tatars. After two days of intoxication the inhabitants became so reckless that they believed the Tatars' promises and opened the gates. The result was that Moscow was sacked and ruined.[40]

In 1433, on the River Klyazma about twenty kilometres from Moscow, Vasily Temny was routed and captured by a small force of soldiers loyal to his uncle Yury of Zvenigorod solely because, the chronicle states, "he received no help from the Muscovites, for many of them were drunk, and had taken mead with them in order to carry on drinking".[41]

It is thus quite clear that in the fourteenth and fifteenth centuries very significant changes were occurring in the way in which intoxicating drinks were produced, and that as a result the character of these drinks was changing substantially: they were becoming more powerful and stupefying.

It is recorded that in 1386 the Genoese ambassador, on a journey from the Genoese colony of Cafta (Theodosia) in the Crimea to Lithuania, brought with him *aqua vitae*,[42] which had been discovered by alchemists in Provence around 1333–4 and which had become known in the South of France and the adjoining parts of northern Italy.[43] This strong liquor was known in the

Tsar's court, but it was considered too powerful for an ordinary drink, suitable only for use as a medicine and then only when diluted with water. It is quite probable that the idea of diluting alcoholic spirit with water gave rise at this time to the Russian modification *vodka*, at least as a term for such a mixture. There was, however, another possible source; in terms of production technique vodka very likely grew out of *sikera* and from the making of *kvas*, and in part also from the making of mead in the form which this had assumed by the thirteenth and fourteenth centuries.

To clarify this question we shall turn to examining the technology of the drinks whose terminology was outlined earlier. But before this, we shall briefly set out the chronology of the rise of all alcoholic beverages from the ninth to the fifteenth centuries.

References to Alcoholic Beverages from the Ninth to the Fourteenth Century

Late 9th century	c. 880–890	matured mead
Early 10th century	907	grape wine
Early 10th century	921	"drunken" birch sap
First half of 10th century	920–930	hopped mead
Late 10th century	996	brewed mead
Late 10th century	988–998	grape wine (sour, sweet)
Mid-11th century	1056-1057	*kvas*
Mid-11th century	1056-1057	*sikera*
Second half of 11th century		"unfinished" *kvas*
First half of 12th century		"prepared" *kvas*
Second half of 12th century		"prepared" beer
Late 13th century	1265-1270	*khmel'noe* (intoxicant)
Late 13th century	1284	*ol* (ale)
13th–14th century		"prepared" wine

From this brief table, summarizing all the previous historico-linguistic analysis of the terminology of alcoholic drinks, two basic conclusions can be drawn.

First, the earliest alcoholic drinks to appear, and those which dominated the scene throughout the entire Old Russian period (that is, from the ninth to the fourteenth centuries) were those which used as their raw materials natural sources of sugar: birch sap, grapes and honey. Of these, grapes and the wine produced from them were of completely foreign origin, and were used only in the highest ranks of feudal society or in religious ritual. However, grape wine served as a standard for alcoholic liquors both in the strength of its effects and in its terminology. Less potent than wine were the weakly alcoholic drinks such as birch sap and *kvas*. Stronger than wine were "drunken" birch sap, *kvas-sikera*, ale, matured mead and brewed mead. The fact that honey and birch sap were the first natural Russian raw materials for the production of alcoholic beverages, and that there was no need to subject them to a complex process – it was enough to let them ferment naturally – for many years made it unnecessary to develop more advanced technology for Russian alcoholic beverages.

Second, it was not until the second half of the eleventh century, or perhaps the end of the eleventh and the beginning of the twelfth century, that there is evidence of the development of alcoholic liquors made from ingredients other than the simple ones mentioned above. Then, the new raw materials were cereal grains. This production undoubtedly developed as a sideline of baking, and was closely linked with the dispute that was raging in the Church at this time over the proper way to celebrate the Eucharist: with unleavened bread or with bread raised by yeast.

The production of grain-based alcoholic beverages, which in the twelfth and thirteenth centuries gave rise to such drinks as beer and ale, was further developed and extended as a result of the Mongol-Tatar invasions, which isolated Russia from Byzantium and led to the transfer of the Russian political centre to the region of the Oka and the upper Volga, where the main food staples were rye, oats, barley and honey. With the loss of the sources of grape wine, Church and state sanctioned the production of alcoholic drinks from grain. The Church was also obliged to grant permission to substitute ale for wine in church ritual.

The appearance around the end of the thirteenth and the beginning of the fourteenth centuries of such terms as *khmel'noe* (intoxicant) and "prepared wine", signifying alcoholic beverages in general, is evidence of the extremely vague standards applied to these drinks. There were no definite names, and the only general features were the use of hops and an alcoholic content similar to that of wine unmixed with water.

The fact that apart from mead all alcoholic drinks – ale, beer and "unfinished" *kvas* – were known as prepared wine is still more evidence that a process was being employed for making alcoholic liquors using a non-honey base; that is, with grains. Such drinks, for which a single, general name still did not exist in the thirteenth and fourteenth centuries, were referred to as *khmel'noe*, since all of them required the use of hops. When hops were used in the production of brewed mead, the resulting drink was known not just as *khmel'noe*, but also as *khmel'ny med* (hopped mead). Thus, *med* remained the only one of the old terms for alcoholic beverages that was not combined and confused with later ones. This explains the fact that the term *med* is well represented in folklore, and that it has links with antiquity, although it only became common in texts in the fourteenth or even fifteenth century.

Another term much favoured in folklore, *zeleno-vino* (herb wine), is completely absent from written sources. Two explanations for its emergence can be given. First, it may have arisen as a metonym of the word *khmel'noe* (hopped, intoxicant); *zeleno* is not identical in meaning to *zelenoe* (green), but is the same as *zel'eno* (greenery, verdure). Thus, the term meant "flavoured with herbs" (such as hops and St John's wort). Second, the term *zeleno-vino* may have arisen later, during the seventeenth or eighteenth centuries, as a reference to the cloudy green tinge of low-quality home-distilled spirits. The first explanation appears more likely. The fact that the term *zeleno-vino* appears in folklore no earlier than that dating from the fifteenth or sixteenth centuries indicates that the term itself could not have arisen before this time. It clearly refers to "grain wine", not to mead; and this testifies indirectly to the fact that the fifteenth century saw the full transition from the earlier alcoholic drinks, whose terminology we have examined, to a new one: grain spirit.

This review of the terminology and of the dates when it appeared and was used shows that in the period from the ninth to the thirteenth centuries a handful of terms, denoting mead, wine, *kvas* and birch sap, were employed in a comparatively stable fashion. During this time their meaning did not change significantly, though it was refined somewhat, and new nuances appeared in the terminology. In the thirteenth and fourteenth centuries new terms for alcoholic beverages came on to the scene. Moreover, historical and economic evidence indicates that by the fifteenth century the supply of honey, the age-old raw material for the production of mead, was diminishing. Then in 1386 the Russians became acquainted with *aqua vitae* brought from Cafta, the Genoese colony in the Crimea. All of this indicates that in the fourteenth and fifteenth centuries a turning-point was reached in the production of alcoholic beverages.

This new historical environment helped bring about changes both in the raw materials and in the production techniques of Russian alcoholic liquors. However, since all this evidence is indirect in nature, it is extremely important to determine whether direct evidence exists of changes in the technology of production. To this end it is necessary to survey the available information on the technology of production of alcoholic liquors in the period before the fifteenth century.

Techniques for the Production of Alcoholic Drinks in Russia in the Fourteenth and Fifteenth Centuries

Earlier it was noted that literary evidence – chronicles, other documentary sources and folklore – provides us with outlines of three methods for producing alcoholic liquors: natural fermentation of fruit juice and tree sap, natural fermentation of honey, and the heating of a mash and its subsequent fermentation.

These three means of obtaining alcoholic beverages are closely linked, having arisen in turn one from another. It is reasonable to suggest that the successive development of these techniques ultimately constituted the source of distilling, which was a completely new method but could never have been discovered without the experience which had already been acquired. If we are to establish when distilling could have arisen, we need a clear conception of all three methods of preparing alcoholic beverages that were in use before distilling appeared.

The first method, involving the spontaneous fermentation and oxidation of the sap of birch trees (or in Belorussia, Volynia, and in Polotsk Russia, of maples) resulted in a brew that was cloudy, contaminated with acids, and not unlike vinegar, but which was intoxicating. This method underwent a refinement which involved the use of grain (usually in the form of flour) instead of birch sap. The cereal was mixed with water and allowed to ferment either naturally or with the help of malt, yeast or fermented dough, whose enzymes turned its starch to sugar. In this way *kvas* was obtained. Since this drink was originally a substitute for birch sap, it was prepared once a year on 1 March; that is, at the New Year (until 1492 the year began in March, another instance in which the fifteenth century formed a turning-point). The mash was prepared in large barrels, and for almost the whole of the rest of the year water was added in quantities equal to the *kvas* taken out.

This later *kvas* had not fermented entirely, and was largely a weak solution of rye malt. For this reason *kvas* and beer were considered to be especially good in March, but after that they became more and more watery. This is the origin of the expression *martovskoe pivo* (March beer) – that is, good, strong, full-flavoured beer.

The second method, *stavlenie* (standing) – that is, the maturing of a sour fermented mash – was developed through the production of matured mead. This method was therefore also called *medostav*. It was used only for honey. It required large quantities of this, and took an extremely long time to yield the finished product (at least five years, but it might be matured for as much as forty), and yielded a relatively small quantity of liquor for the quantity of ingredients used. By the fifteenth century "standing" mead was already regarded as a luxury, and was comparatively rare.

Although V.I. Dal in his dictionary records the word *medostav*, he explains it incorrectly, identifying it with the word *medovar* (mead-brewer). He misses the point that what was involved here was not a profession but a method, an area of technology, since *medostav* signified the maturing (*stavlenie*) of mead. The word *medostav* is also used for a person, but only for someone engaged in the maturing of mead; an individual quite distinct from a *medovar*, who made the cheaper brewed mead.

As far as can be judged from the later account in the work on housekeeping, the *Domostroi*, the process of *medostav* took place as follows. Sour mead and berries were placed in a cauldron, so that the berries began to ferment. After this the berries and the mead were heated, so that the berries rose to the surface, and the mixture was thoroughly cooked. Then it was left overnight to allow the dregs to settle. Next day the mead was poured off (without being strained) into casks, which were coated with pitch and placed in a cellar to mature for several years. Sometimes the casks were buried in the earth, which prevented air from entering but allowed gases to escape.

The preparation of the intermediate product, sour mead, required considerable advance notice and was extremely labour-intensive. First the honey was dissolved in water, which was added in the proportion of four or six parts to one of honey. Then the wax was skimmed off the surface, and the solution was strained through a sieve. A measure of hops (or a half or quarter measure)[44] was added for every *pud* of honey, and the solution was boiled down to half of its volume, while the froth was constantly skimmed from the surface. After boiling the mixture was allowed to cool, and yeast was added (or sometimes a piece of warm, incompletely baked rye bread), together with *patoka*, a syrup derived from starch. The brew was then placed in a warm stove with the fire raked out, so that the mead cooled slowly. It fermented but did not turn completely sour. With great care being taken to pick the right moment, the mead was then poured off into barrels which were placed on ice, delaying further fermentation. The sour mead was then used as an ingredient for the production of various types of mead – boyar's mead, syrup mead, mead from scalded honey, white mead, herb (that is, spiced) mead, and various berry meads (cranberry, bilberry and blackcurrant). The method was connected with one more rare and now vanished product: isinglass, prepared from the swim bladders of fish, usually sturgeon. This was added to the prepared mead before the casks were coated with pitch, in order to retard the process of fermentation and to neutralize the waste products that built up in the mead.[45] Since isinglass was vastly more expensive even than sturgeon caviare (in the first half of the nineteenth century a *pud* of caviare cost 15 roubles, while a *pud* of isinglass cost 370 roubles), this further increased the price of matured mead. The length of the process of *medostav*, the high cost of production and its insignificant yield led eventually to the technique giving way to the more profitable *medovarenie* (mead brewing).

Mead brewing developed virtually in parallel with *medostav*, but its importance grew especially quickly from the thirteenth and fourteenth centuries. This method for the production of alcohol from honey became the usual one, and in the course of the fifteenth century its dominance became absolute.

Mead brewing differed from *medostav* first of all in that it used a more dilute honey solution: the quantity of water was seven times that of honey. Starch syrup was added to raise the percentage of sugar in the mash and hence increase the eventul alcoholic content. In some cases another ingredient was added to the mead: beer mash, resulting in a product that was no longer brewed mead but beer laced with mead. However, the basic technical novelty, the distinguishing characteristic of brewed mead, lay in the adding of yeast to the honey solution, together with the fact that when the mixture was boiled before being allowed to ferment, water in which hops had been boiled was added.

The yeast was added after this boiling, and the brewed mead was decanted several times into different barrels in order to stop the fermentation process at a particular stage. The pitch-coated barrels containing the mead were put on ice after the second or third decanting; the fermentation in them was considered to have been "frozen" to the extent that its products were not transformed into gases, but entered the mead itself, imparting to it a special power of knocking drinkers off their feet.

The technology of mead-brewing grew increasingly sophisticated and served as a model for the brewing of *kvas* and beer. Techniques underwent especially rapid development in the thirteenth century, reaching great refinement in the production of the high-quality *ol* (ale).

The brewing of beer, especially of superior varieties, to some extent also possessed a ritual character. For these products beer-brewing remained an extremely costly and, most importantly, extremely labour-intensive procedure demanding great quantities of raw materials, huge cauldrons, a brewery building and a team of workers. Such beer was brewed for several hundred people at once, and needed to be drunk as quickly as possible, in the space of two or three days, since it could not be stored.

The brewing of beer could not develop into a regular industry, since it was done only for festivals. This, more than anything, limited the frequency with which beer was prepared – there were only two or three such occasions in the year.

Mead-brewing, which was more independent of the time of year, and which gave a more highly alcoholic liquor which kept for longer, remained the predominant method of preparing alcohol right up to the fifteenth century. This explains why the techniques of mead-brewing served as the basis for the production of alcoholic beverages from other raw materials. The technical experience of mead-brewing was transferred to the brewing of beer and *kvas*, and it was out of this that distilling then arose.

However, there was nothing instantaneous about the rise of distilling. It developed slowly until economic, and to some degree sociopolitical, conditions made it necessary to replace the old, wasteful methods of producing alcoholic liquors with new, more efficient ones.

It appears that distilling or, to be more precise, the first steps towards it, originated at the end of the twelfth and the beginning of the thirteenth century. This occurred among the poorest members of the agrarian population, as part of the process of development of *kvas*-brewing. But since the Tatar-Mongol invasion began around 1230, throwing the entire political and economic life of Russia into turmoil, this transition to distilling was cut short, and *kvas* brewing remained at its existing level for another two centuries.

The making of *kvas* in Russia involved keeping the wort in open containers for at least a year. Sometimes, especially among the poor, the new wort was poured into the old container without cleaning the vessel or even emptying out the remnants of the previous year's wort. This created a long-lived culture of micro-organisms which, it was noted, improved the quality of the product, and increased its alcoholic content. A further peculiarity of *kvas*-making was that, in contrast to the practice of brewing in Western Europe, the wort was prepared not from whole cereal grains but from coarse flour, which might be a by-product of the preparation of fine flour for bread.

This practice of making *kvas* and beer out of the wastes from breadmaking – and indeed from everything which could be scraped together in the granaries – gave rise to a third peculiarity of the production of alcohol in Russia: the raw material did not consist solely of a particular type of grain (for example, barley, as in Britain), but of all types or, more precisely, of all the types of grain waste that were available. This led to the appearance of a *kvas* mash prepared from a mixture of rye, oat and barley flour, and sometimes also buckwheat. It was observed that the mixture of various types of flour gave a stronger and tastier product than flour of a single type. Brewers began consciously to select the proportions of these three or four types of flour, seeking the most suitable combination. Through trial and error it was discovered that of the various cereals, rye was the best for producing both kvas and beer. While continuing to use mixtures of various types of flour, brewers also, when they had the chance, began to use rye alone as their raw material.

A further peculiarity of the making of *kvas* and mead in Russia was the widespread practice of adding herbs and spices. The principal addition was of hops, but other plants were also added: St John's wort, wormwood and caraway. The aim was first of all to increase the storage life of weakly alcoholic drinks by preventing them from turning sour; this was becoming especially important as the volume of their production increased toward the end of the fourteenth century. Second, the additives partly masked the unwanted flavours that resulted from the shortcomings of the preparation process. Third, hops imparted a characteristic odour to all alcoholic products, giving them a standard, recognizable quality, something which was extremely important for the establishment of a state monopoly in alcoholic beverages.

Hops were known in Russia from the tenth century; the word *khmel'* is encountered in Nestor's chronicle. By the thirteenth century their production had expanded to the point where the city of Novgorod was buying them in large quantities in the principality of Tver and exporting them to the German states through Riga. Merchants from the Hanseatic cities bought hops both from Novgorod and from the principality of Smolensk, and in 1330 hops were mentioned as an item of export in the treaty between Polotsk and Riga.

The evidence of economic history thus confirms that a boom in the production and use of hops occurred in the thirteenth and fourteenth centuries. It is important to stress that while hops were used in the German brewing industry mainly to clarify beer, and the quantities in which they were used were therefore relatively small, in Russia they had other functions. They were used not so much in the making of beer as in the production of mead, the brewing of *braga* (light beer), the making of *kvas*, and in the distilling that was then in its formative stages. As a result hops were used in greater quantities than in Western European beer-brewing.

Moreover, the hops were added not at the end of the brewing process but in the middle; one could even say almost at the beginning, together with boiling water. The active ingredients were extracted through boiling or, in the production of matured mead, by heating the hops together with honey.

The production of alcoholic liquors in Russia was thus distinguished by the use of additional vegetable ingredients – in the case of hops, in lavish quantities. The main raw materials also had a special character: they consisted of honey and of rye, oat and barley flour, while the malt was always made from rye and not barley. The non-vegetable raw materials included isinglass, which was used on a much larger scale than in other countries.

The technology employed differed from then current practice in Western Europe in various ways: the prolonged, gradual fermentation of the mash; the abrupt addition to it of boiling water, the subsequent malting and fermentation (with malt, leaven and yeast), and finally, the use of "stewing" in a falling temperature to bring about a slow secondary fermentation.

It was usual elsewhere to boil the mash for a prolonged period; but the Russian practice was not so much to boil it as to pour boiling water over it. This ensured the slowness of all the subsequent processes. Although it retarded the pace and volume of production, it made possible a product of uniquely high quality, as long as the process was not hurried. As soon as the brewers tried to speed up the process, the quality of the product became extremely poor. This is obviously how we should understand the term "unfinished" as applied to *kvas*.

Cases of drinkers being poisoned by "unfinished" *kvas* in all probability induced brewers to boil the mash and not simply to pour boiling water over it. It seems likely that this new practice was adopted in the thirteenth century after the Tatar-Mongol invasions, when brewers realized from the example of the Tatar *katik* that fermentation could also occur in a boiled mash; though this required a greater quantity of yeast of a different, more potent and more durable kind, and different temperature conditions for fermentation.

It is possible that the practice of boiling the mash had already begun before the thirteenth century, but that it was only many decades later, probably in the course of heating the malted flour mash, that the accidental distillation of alcohol occurred. When and how this happened we can only guess. But a survey of archaeological evidence and a comparison with the development of other forms of production employing heating as a basic method will help establish what the technical preconditions were for this to occur.

Let us begin with other forms of production. The one which most closely resembles the distillation of alcohol in the type of equipment and technology employed is the extraction of pitch. Here even the terms are similar. They must have been even closer in the remote past, when people used the expressions *sidet' vino* (to "sit" wine), and *sidet' smolu* (to "sit" tar).

Tar extraction in Russia first acquired importance in Polotsk and Novgorod, and in the northern part of the princedom of Smolensk (the very name of the town Smolensk comes from the word *smola*, pitch). The rise of the industry was prompted by two factors. One was the development of traffic on the rivers and lakes of the plain, connected by the powerful Western Dvina river with the Baltic regions and the sea. Here the most rapid and reliable means of communication and of transporting goods was by water. Boatbuilding required a substance to waterproof wood and fill seams. The second factor was the infertility of the boggy, forested region, which made it impossible for the population to produce enough grain for its needs. The inhabitants therefore sought to obtain grain in exchange for the products of three local industries: fishing, the collection of honey from wild bees, and the extraction of pitch. The latter, however, is recorded for the first time only at the beginning of the seventeenth century. At that time pitch was referred to in Russian documents as *smolchug*. The modern word for pitch, *degot'*, appeared only later; it was a borrowing from Lithuanian, in which it means "black

tar". The principal areas of pitch and tar production were Polotsk
Russia, Belorussia and Lithuania.

The fact that the word *smolchug* first appears in official state
trade documents only in 1617, in connection with the Peace of
Stolbovsk, does not of course mean that pitch and its production
were unknown before this time. In this connection one should
recall a most important observation by A. Shletser:

> Unfortunately, the Russian chroniclers of trade affairs left behind an
> incomparably poorer heritage than all the other chroniclers of the
> medieval period. If Russians want to learn about the history of their
> trade during those times they must seek it among foreigners, whose
> writings, also meagre and often confused, have been gathered
> together only very recently. . . . The chroniclers do not say a word
> even about the famous Hanseatic office in Novgorod, which the
> Hanseatic merchants themselves considered the most important of all
> their offices.[46]

The first mention of domestic trade in pitch is found under a
chronicle entry for 1264, and of foreign trade under one for
1373 – that is, in the late fourteenth century, when this trade
was already flourishing. Hence we have to react sceptically to the
first mention of pitch in the trade documents, and to recognize
that this does not relate to the beginnings of pitch extraction,
only constituting evidence that it had reached a significant level
of development. The first mention of pitch in everyday documents
of the Moscow state is considerably earlier: the word *degot'*
occurs in a source from 1568, and the phrase *dekhtrianaia iama*
(pitch pit)[47] is found in one from 1517. In north-eastern Russia
such references are found even in a document from the years
1496 to 1498: ". . . and, lord, they cut down the forest and
stripped the birch trees for pitch".[48] There is thus a discrepancy of
two centuries compared with the trade documents. But the date
is still no earlier than the fifteenth century, and the end of the fifteenth century at that.

Here we are again struck by a nearly complete analogy with
the production of alcoholic beverages. First of all, the fifteenth
century appears as the critical period; during this time – it might
have been at the beginning, the middle or the end, but no earlier
– the heating of wood began to be used as a method of produc-

ing pitch. It seems no accident that the distillation of alcohol probably arose during the same century.

At the same time there is no doubt that pitch was used for shipbuilding and other purposes much earlier than the fifteenth century, since the name Smolensk dates from the eleventh century. From this we can only conclude that what people originally understood by the word *smola* was indeed pitch; that is, the resinous exudations of coniferous trees. This substance was used extensively; it was gathered (like honey, in the forests), heated (like mead) in cauldrons, and in a hot, molten state was smeared over boats, barrels and the foundations of buildings. Pitch was obtained by heating the pinewood which contained it. Later, it was discovered that birch trees, especially their bark, also yielded pitch.

One way of extracting pitch was by boiling it out of pine logs in water. Another was the "sitting" (*sidka*) of pitch; that is, dry extraction in pits, with channels at the bottom to drain off the product into a reservoir. There were channels at the top of the vat to catch the pitch which evaporated in the boiling process. In these channels, the pitch cooled and condensed to a liquid so that it could be collected. It was these channels which gave rise to the idea of pipes – closed channels – in the distillation of alcohol, where they caught the alcohol vapour as it boiled off at a temperature below the boiling point of water.

In this way the extraction of pitch prompted the idea of the distillation of alcohol. The ideas of pipes and of cooling could not have arisen spontaneously out of the making of beer or mead, but they were quite natural and even inevitable in the case of extracting pitch. Here the new product could not be collected at all unless it was cooled, since vapour would simply disperse. It was no accident that pitch was also known as *var*, since in Old Slavonic this word meant "great heat" or "boiling water".

There were other technical analogies as well. The heating of resin gathered from the surface of trees, which took place in the period from the ninth to the twelfth centuries, corresponded to the brewing of mead from honey gathered in the forests.

Pitch extraction, and especially the "sitting" of pitch by the dry method, was a quite new method of production, the obtaining of an extract, and was fully comparable in historical and technical terms to alcohol distillation, which involved obtaining an extract from a grain mash.

Even the fact that originally both pitch extraction and alcohol distillation were described by the same term – *sidenie* or *sidka*; that is, slow "sitting out", patient and uninterrupted production – indicates that in historical terms the extraction of pitch did not simply coincide with the distillation of alcohol, but preceded it. This is apparent because a slow "sitting out", involving gradual heating, was essential for the production of pitch. In the distilling of alcohol this requirement was not so obvious, since there the materials involved were different – both the mash and the product of distillation were liquids – but overheating, which allowed the water in the mixture to boil and pass into the pipes, would spoil the result. Meanwhile the first instructions for distilling alcohol that have come down to us contain abundant advice that the process should go ahead as slowly as possible. This text reads very much like instructions for producing pitch: do not allow the process to be interrupted, but build up the heat slowly, while all the time sitting and tending the fire. In just the same way the use of the verb *gnat'* – to "drive out" pitch or alcoholic spirits – testifies to the fact that in technological terms the extraction of pitch preceded the distilling of alcohol. The point is that it was necessary in the full sense of the word to "drive" pitch out of the wood in which it was contained, and moreover, to drive it out slowly, persistently, over a long period – that is, to "sit it out".

In pitch extraction the terms used have obvious meanings. "Driving out" and "sitting out" characterize the method of production precisely.

In alcohol distilling the same words are almost devoid of sense. They are used here as technical terms pure and simple. This proves better than anything that they were borrowed from another, related process of production. Such terms as "to sit out" do not in any way flow from the malt or mash that in the process of alcohol distillation is heated or boiled. These latter terms recall more than anything the cooking of food such as porridge or cabbage soup, the more so since their purpose was to bring the mash more quickly to a vaporous state. At first glance the term "to drive out" makes sense, since the alcohol is indeed driven out of the mash. But although this meaning is obvious to us today, it was by no means clear at the time when the method of distilling alcohol was first developed and used. Consequently, the term "to drive out" could not have arisen in this way. Its origins lie in the extraction of pitch, just as the channels in pitch extracting inspired the idea of pipes for distilling alcohol; though it must have taken time to realize that a closed pipe was necessary to catch light alcohol vapour instead of the heavy, easily condensed pitch.

But how, then, are we to explain the quotation, coming from an eleventh-century source and cited in the terminological section of this study, which warns of "woe to those who chase after kvas" (kvas goniashchim)? Does the use of this latter phrase not prove that people at this time were producing grain spirit, or to put it more simply, something like vodka? This extremely enticing and apparently credible suggestion falls down, however, when we note that in Old Slavonic the verb goniti, from which the participle goniashchii comes, meant to pursue, or to desire or crave. In the case in question the verb is used in the second sense, and the significance is that "he who craves kvas, sikera and other alcoholic drinks" calls down misery on himself. The admonition is thus directed against drunkenness, and tells us nothing about the technology of production of alcoholic liquors at that time. It does not even hint to us what the nature of that production might have been, since the verb gnat' did not yet exist, and the verb goniti signified an abstract concept rather than a concrete one that can be related to production.

In this way an examination of the technological process also leads us inexorably to the conclusion that the distillation of alcohol could not have spontaneously arisen before the middle of the fifteenth century, even if we acknowledge that pitch extraction was known from the beginning of that century.

To obtain a fuller picture of the technology of production of alcoholic liquors in Old Russia and to arrive at more soundly based conclusions on when the distillation of alcohol could have appeared, we shall now turn from the written sources in the linguistic, philological and historico-economic fields to technical sources. In other words, we shall acquaint ourselves with the equipment and techniques which were used in the production of kvas and beer, and try to ascertain whether this equipment could have enabled people to produce grain spirit.

The Earliest Forms of Technical Equipment before the Rise of Vodka Production

The technical equipment associated with the earliest methods of preparing alcoholic beverages was simple in the extreme. At first there was simply a barrel or a large earthenware vessel, simply a container for liquids made out of whatever material was characteristic of the region. In north-eastern Russia this material was wood, and in southern Russia and the Ukraine clay. Here we find a close analogy with the countries where grape wine or fermented milk drinks had been produced since ancient times. In these countries the most suitable materials available were either the skins of domestic animals, or ceramics. Hence both wine and *kumys* were produced and stored in goatskins, sheepskins or horseskins (*burdiuki, torsyki, saby*) or in earthenware pots (*kuvshiny*). All of these words in the Iranian, Turkish and Georgian languages have the meaning of vessels or containers, like the word *keramion* in Greek, which is derived from the word for clay.

In these containers alcoholic beverages could only be fermented and stored; brewing in the sense in which we now understand it – heating the ingredients over a fire – was impossible on anything more than a small scale. For this a metal vessel was essential. In Old Russia before the Mongol-Tatar invasion such containers were extremely rare, apart from gold, silver, copper and bronze vessels that were used in religious rituals. Nevertheless, mead was prepared in metal vats. Beer and *kvas* were brewed in ceramic pots. However, the process of brewing in earthenware vessels was somewhat unusual, and therefore the expression *korchazhnoe pivo*, "pot beer" – that is, home-made beer brewed in a pot heated in a Russian stove – has come down to us. The term "pot mead" also denoted mead prepared in a stove. The word *korchaga* (pot, jug) is of extremely ancient origin. It is encountered in Russian chronicles for the year 997; that is, the same year as the first mention of brewed mead.[49] In Old Slavonic the word *kr"chag"*, from the Greek *keramion* (vessel), meant a water jug. The Old Russian word *k"rchaga* evidently came directly from the Turkic *kurchuk* (vessel), an even older word than the Greek one.

By the eleventh century the word *korchaga* had already acquired a distinct meaning, that of a large, open earthenware vessel with a wide neck and narrow base. Russian cast-iron stoves were later made to designs similar to these *korchagi*.

Such *korchagi* have been found in great quantities in excavations in Novgorod Russia, especially in the regions of Vladimir and Suzdal, in strata of the eleventh, twelfth and thirteenth centuries. These ancient vessels are virtually identical in shape to pots made in the nineteenth century, though the latter are smaller. Russian ceramic vessels shaped differently from *korchagi* have different names: *molostovy*, *krynki* and *gorshki*.

In the Ukrainian language at the end of the nineteenth century the word *korchaga* meant a narrow-necked container for vodka.[50] Two striking points are evident here. Firstly, the form of the objects denoted by the term *korchaga* in Russian and Ukrainian was in direct contrast. Secondly, the meanings coincided in the area of use to which the vessels were put. The precise manner of use was different; in Russian *korchaga* denoted a vessel for the production of the drink; in Ukrainian, one for the storage of the prepared product ready for consumption. But despite these apparently sharp differences, through all the contrasts in the meaning of the term and in the external appearance of the objects which it designated, the term in itself was firmly linked to alcoholic beverages: *korchazhnoe pivo* (pot beer), *korchazhnyi med* (pot mead), a *korchaga* for the brewing of beer (and later, for spirits), and a *korchaga* for storing vodka. If we are to consider that the word *korchma* (tavern) was also derived from the word *korchaga* (*korchma* used to denote a place where pot beer was brewed and sold),[51] then it is entirely valid to suggest that *korchaga* itself originally denoted a vessel used in alcohol distilling, that is, in the use of heat to drive off alcohol from the cereal mash. However, this "pot distillation" was an initial experimental process which was not developed further once the large-scale production of grain spirit got under way. As we know from literary and historical sources, this occurred in the sixteenth century, or possibly even somewhat earlier, though there is no precise evidence on this. But there can be no doubt that originally alcohol distilling involved the use of earthenware pots, or that this process was linked with the peculiarities of Russian technology; that is, with the availability of such equipment as the pot, the tub, the basket and the Russian stove.

Beer and mead were "sat out" in the following fashion. The fermented beer wort or honey solution was poured into pots, and the latter were placed in the stove and covered with other pots, so that the wort heated up. Meanwhile, in order to avoid wasting wort, and in case the hot liquid boiled over, a wooden tub was placed beneath. It is possible that during a long period in the stove, where there was an even temperature, the brewing was accompanied by spontaneous distillation, the products of which condensed and dripped into the tub. If the beer wort were replaced with coarser products – oat, barley and rye flour – and if this mash were "sat out", grain spirit would have been obtained. It is true that this would have been very weak, but it would have been capable of inspiring the idea of improving the technology and of transforming the process into one of regular distillation.

Various proverbs and sayings, now forgotten and incomprehensible, hint at these processes; they include "Oh, you are a tub covered with a pot", and "Happiness is a tub covered with a pot". The reference in the first of these sayings was either to a *korchmar*, the proprietor of a *korchma* and the brewer of pot beer, or to a drunkard. In time these associations became automatic, and the original meaning of the proverb was lost. The second proverb declared that happiness lay in drink – that is, in vodka. We find an echo of this proverb in Pushkin's *Tale of the Fisherman and the Fish*, where a broken tub serves as a symbol of unhappiness. The original, ancient form of this symbol of happiness, the tub – not just any tub, but a tub covered with a pot – was perceived even at the beginning of the nineteenth century as a synonym for vodka, though by this time very dimly. This indicates once again that pot distilling, if indeed it represented the very earliest stages of development of the production of alcoholic spirits, was an isolated phenomenon and was quickly replaced by a better process, having passed on only the idea of distillation. We can surmise that pot distilling may have occurred even before the fifteenth century.

As a seemingly inevitable sideline to the production of beer and mead, pot distilling was also known in the Ukraine or, more precisely, in the territory which did not become part of northeastern Russia after the Tatar-Mongol invasion. The Ukrainian word *makitra* provides indirect evidence of this. It denotes the same vessel as the Russian *korchaga*. It retained both its form and its function, but changed its name. The word had strong links with the concept of alcoholic liquors, even though contemporary *makitry* are used for storing dry goods. Hence we find the expression "a head like a *makitra*", which means that someone has a severe hangover. In addition, the verb *makitrit'sia* means "to feel one's head spin", while *svit makitrit'sia* refers to one's head "turning around". Thus in the Ukrainian language the word *makitra* and other words with the sense of "pot" are linked with ideas which suggest above all an alcoholic drink such as grain spirit rather than mead, which does not leave one with a particularly bad headache. Here as well we have a weak echo of ancient impressions of a drink whose low quality was the result of pot methods of production. This drink must have been the original, unrefined grain spirit or "unfinished" beer.

The point is that with pot production, involving the extended "sitting out" of cereal ingredients, a genuine separation of the alcoholic products from the watery elements of the wort did not occur. Most of the liquid remained in the vessel, and the main effect of heating was to concentrate it by driving off water – and also the ethyl alcohol, the ordinary alcohol found in properly distilled spirit, and which boils at a lower temperature than water. The alcohols which remained and were concentrated were the heavier congeners such as butyl and iso-amyl alcohol, poisonous and foul-smelling, forming the substance known as "fusel oil". There would be other impurities: the composition of the original liquid was uncertain because of the variability of the grain used, and the timing and temperature of the process were ill controlled. Hence, if the production of grain spirit was possible by the pot method earlier than the fifteenth century, it was not developed, since it was expensive, slow, and pointless given the availability of good-quality mead, ale, and grape wine.

This review of the chronology of the emergence of various alcoholic drinks, and its comparison with the available evidence on the equipment used for alcohol production from the eleventh to the fourteenth centuries, allow us to assert definitely that at least until the fifteenth century alcohol distilling – that is, the production of grain spirit – was not known in Russia, whether in

Kievan Rus or Novgorod, or in Vladimir and Suzdal. All these
territories were rich in the natural raw materials (honey and
berries) for making mead, and grain was used there for making
drinks of the beer type – *kvas* and brewed beer – but with a high
content of sugar, malt and hops. It was out of the production of
these drinks that alcohol distilling arose, passing initially through
a stage of pot distillation. This, however, could not have happened earlier than the fifteenth century.

Hence we do not have clear, precise and incontrovertible documentary evidence of the production of grain spirit before the
beginning of the fifteenth century. On the contrary, all of the linguistic, technical, historical and economic materials available to
us indicate that at this time vodka had not yet been discovered,
and that alcohol distilling did not yet exist in the region.

2
Vodka from the Fourteenth to the Nineteenth Century

The Rise of Distillation

In the previous section we concluded from an analysis of linguistic and chronological evidence, and also from a study of what is known about the technical and productive base of alcohol distilling, that the advent of a grain spirit, a genuinely new product, could not have occurred earlier than the second half of the fourteenth century or the first half of the fifteenth. It is known with almost complete certainty that at the beginning of the sixteenth century, around 1505 to 1510, alcohol distilling in Russia was relatively developed. Vodka thus made its appearance at some point during the intervening period of a century or a century and a half. It is, therefore, this period in which we need to conduct our search for a more precise date for the first production of grain spirit in Moscow state or in the other states making up Russia at that time – the principalities of Tver, Nizhny Novgorod and Ryazan, and the republic of Novgorod.

It should not, of course, be imagined that the advent of alcohol distillation, which had a long prehistory in *kvas* production, and which was the outcome of the whole preceding course of historical development of alcoholic beverages, could have happened all at once. There would have been a gradual transition, taking place over many years, from the making of *kvas* and beer to pot distilling, and after that to genuine distillation with its specific equipment and techniques.

However, it is entirely possible and necessary to specify, if not a particular year, then at least a period of ten or twenty years in which the rise of alcohol distilling in Russia is most likely to have happened. This is feasible because historians have at their disposal not only direct historical sources such as printed texts and other material evidence, but a whole arsenal of indirect means of arriving at historical proof: notably, economic and social data. These can cast light on when vodka first appeared.

Earlier it was stressed that vodka is a special product, in the sense that it is closely linked with state finances and the social problems of various societies and social formations. It follows that the search for a date for the emergence of vodka must be conducted through a detailed, careful analysis of the economic and social conditions existing in the state.

In other words, we must look closely at the social situation during the period from the middle of the fourteenth to the beginning of the sixteenth century, combing through all the events of these 150 years. We must ask whether there were moments when the situation which had existed earlier altered sharply; whether there were leaps, shifts, or other changes in the conditions of social life and the life of the state. Any such development during the period would have to be linked in one way or another to the advent of alcohol distilling and of vodka, since only a definite economic shift could call into being a new type of production, while the emergence of vodka would undoubtedly be reflected in corresponding social changes. The timing of economic and social developments will serve to direct us towards a more precise dating of the appearance of alcohol distilling on the Russian scene.

Before we begin to search on historical lines in this way we must, however, define more precisely the character of the economic and social developments in question. What are the features which will enable us to recognize these developments as signposts to the history of alcohol distilling and of vodka?

Factors Expediting or Signalling the Advent of Vodka

There is one crucial economic phenomenon which provides an especially precise pointer to the existence in a country of alcohol distilling as a more or less organized and regular activity. This is a sharp change in tax policy and the taxation system as a result of the introduction of a new factor in state finances: a monopoly of alcoholic spirits, including as a rule both their production and their sale.

The cereals on which the production of grain alcohol is based formed the criterion of value underlying the economy of every medieval feudal state. Consequently, as soon as grain spirit appeared it immediately became the object of the rulers' attention and was made a state monopoly. Such a development was especially likely in the Russian feudal state, which had a mainly agricultural economy dominated by cereal growing. Once vodka had begun to appear on the market, it promptly acquired the character of a concentrated, portable expression of the value of grain. The attention which was paid to it not only by the organs of state finance, but also by private producers and traders, grew by leaps and bounds.

This gives us the possibility of dating the rise of alcohol distilling almost to the year and month, on the basis of the date of introduction of the monopoly on sales of spirits. None of the other alcoholic beverages noted in the chronological and terminological lists in Chapter 1 had any financial constraint placed on it by the state. Neither Old Russia, nor ancient Greece, nor the ancient states of the Caucasus and Asia Minor, imposed taxes on alcoholic drinks; nor, at the other end of the known world, did England or Scotland. From the earliest times the manufacture of wine, mead and beer had been domestic or communal in nature, and had been closely linked with religion and with rituals which can be traced back to the pagan cult of ancestor worship and the accompanying belief in the afterlife. The production of these drinks thus had links with primitive ideology, with sacred and exalted matters. From time immemorial alcohol had been used for important state, festive and religious-political purposes (for example, funeral feasts, annual celebrations marking such events as the sowing and harvesting of crops, and the celebration of military victories). Hence these drinks were regarded not simply as intoxicating, but more importantly as sacred, both because of the way they were used and because of their antiquity. By tradition, therefore, these drinks were not associated with the fiscal interests of the state, which had arisen later and which implicitly recognized the inviolability of the drinks as gifts of nature (based as they were on natural products such as grapes, honey and hops), upon which society could not encroach because of age-old customs inherited from the epoch of the tribal system.

These drinks, such as the palm wine of the Egyptians, the matured mead of the Slavs and the heather-flower *ell* of the Picts, had their origins in the epoch of hunting and gathering long before the appearance of agriculture and pastoralism. Their domestic, familial, tribal or communal, handicraft – but, whatever the case, feudal – character of production rendered them sacred institutions in any state and under any social system which succeeded tribalism, including both the slave-holding system and feudalism.

The distilling of alcohol was a totally different matter. It was one of the first technological processes which was discovered and perfected under feudal society and which went on to acquire a broad significance for society and the state. It arose on a social base marked by the transition from patriarchal feudalism to a money economy, and itself served to open up an easier path to the new economic epoch. This is why the monarchic state, acting in an exceedingly jealous and mistrustful fashion, immediately declared alcohol distilling to be its property, its monopoly.

Winemaking, the brewing of beer and the making of mead always remained free of taxes. These drinks were subject to customs duty only when transported across a frontier, and then at a relatively low rate as freight, in the same way as other goods. But grain spirit immediately became subject to a special tax, almost literally the day after distilling began.

In place of the free, unregulated and unlimited production of grape wine, birch sap wine, *kvas*, mead, and home-made beer and ale, there was suddenly a cruel, pitiless, scrupulously implemented imposition of state authority over the production of grain spirit. The distilling of grain spirit in the form of illicit hooch or *korchma*, as it was known at that time, came to be considered a violation of a vitally important state privilege, in the fifteenth century as in the nineteenth and twentieth.

All this followed as the inevitable consequence of a further social role which alcohol distilling played in contrast to the production of other drinks. Beer, for example, was for centuries prepared by various peoples only at fixed and infrequent intervals, and demanded the labour and material contributions of all members of the community: in Old Russian cities beer was brewed by the whole street or other neighbourhood unit, and in villages by the population as a whole.

For centuries, therefore, the brewing of beer remained an isolated, episodic event, involving distinct communities and groups of people, but not in any way touching on society or social production as a whole, and much less affecting the state. The situation was similar with mead, which was viewed still more clearly as patriarchal, sacred, and a gift from the gods (that is, from nature). It was a gift not simply to the people in general, but to particular groups such as the family, the community, the clan and the tribe; evidence of this is provided by the assignation to specific family and tribal groups of territories where there were wild bees.[1]

Even winemaking, despite its spread through various countries and the effect it exerted on domestic and foreign trade (for example, in France, Italy and Spain), remained in the final analysis a private affair involving individual producers, winemaking firms, and large and small wine merchants. Consequently alcohol distilling, simply by appearing on the scene, brought about changes which, if not revolutionary, at least amounted to a noticeable turning-point in the economy and in society as a whole.

For most countries we can establish the date when it arose with considerable precision simply on the basis of the date when the monopoly was decreed. In the case of Russia, however, documents relating to the vodka monopoly, in common with other documentation on the economy during the fourteenth and fifteenth centuries, have not been preserved. For this reason we cannot cite a particular document, law or instruction to establish the date when the monopoly on alcoholic spirits was proclaimed, but must rely on purely historical research, forming our conclusions from analysis of the changes which would reflect such a development. We need to look for evidence of an abrupt increase in the area sown to crops, and in particular to cereals; for evidence of a significant increase in harvests; for an obvious jump in the volume of trade; for signs of a clear increase in the need for money; for evidence of a switch to commodity-money transactions or of an abrupt broadening of the scale of such transactions in the internal market.

Such a sharp transition in economic conditions is especially characteristic of the emergence of alcohol distilling. The decision to establish a state monopoly on the production and sale of vodka, as with state monopolies on salt and tea, is explained by the fact that in the production of all these goods there is a sharp contrast between the low cost of the raw materials and the relatively high retail price of the final product. This difference remains when one compares the price of vodka with the price of other alcoholic beverages, including even grape wine, and is especially obvious if one considers the yield of the final product in relation to the quantity of raw materials. In the production of matured mead, as we saw in the first part of this work, the volume of ingredients expended in preparing the drink was several times greater than the volume of the final product. Furthermore, it was necessary to use, in addition to the basic raw materials, a number of other high-priced products. The process itself involved high costs, since it took years to complete.

With alcohol distilling everything was reversed. The ingredients were extremely cheap, and the quantities required relatively small; the value of the final product exceeded the cost of the raw materials tens or hundreds of times. If to this one adds the convenience and the relatively low cost of transporting vodka compared to that of transporting grain; the concentration of great value in a small volume; the ease with which vodka could be divided up and sold; and the complete absence of storage problems, since the product did not spoil, it becomes evident why vodka was the ideal product for a state monopoly. In other words, if vodka had not existed, it would have been necessary to invent it, not from any need for a new drink but as the ideal vehicle for indirect taxation.

During the period when the centralized state was coming into being the authorities had an urgent need to secure revenues. It was natural that they should have regarded the introduction of a monopoly on the production and sale of grain spirit as the best way of obtaining the necessary income.

Equally, if circumstances demanded that a way be quickly found to replenish the state finances, this could have speeded the appearance of alcohol distilling, provided that the necessary conditions had been prepared by earlier historical developments. This was precisely where things stood in Russia. The decline by the fifteenth century in supplies of the time-honoured natural raw material, wild honey, that had been used to produce the main alcoholic drinks in Russia, matured and brewed mead, and the development by the end of the fourteenth or the beginning of the fifteenth century of pitch extraction, had already established the economic and technical preconditions for the rise of alcohol distilling.

On their own, however, these preconditions were insufficient. What was still required was an acute need on the part of society and the state for a new source of income, and also a significant general change in historical conditions, such as might have created the need for large sums of money. What was needed, ultimately, was a goal for which large capital investments were indispensable.

In this epoch, the demand by the state for large revenues was created by war, in the first instance by a war of liberation from the Tatar yoke, and also by wars to subjugate the separate principalities and large feudal states – Tver, Ryazan and Novgorod – to the Moscow state; a war to achieve an opening on to the Baltic through taking possession of Livonia with its ports of Narva and Riga; costly defensive wars against Lithuania, attacking from the west; and finally, internal war against feudal boyars in the Moscow principality, to quash their claims to the throne and their fissiparous tendencies. Not least, large sums were required to establish an apparatus of armed force, loyal to the Grand Prince, in the embryonic centralized state. Beside wars there were other goals: the settlement of lands to the south of the Oka and to the east of the Volga, the strengthening of state finances, and spending to maintain the prestige of the Grand Princes. In short, at the end of the fourteenth century there was no lack of opportunity for public spending.

As we search for historical evidence of changes that can be linked to the advent of vodka, we should also bear in mind the extremely important fact that vodka was one of the first newly discovered industrial products to appear in medieval Russia. Gunpowder and firearms, although they appeared in Russia almost simultaneously with vodka, represented a more complex technical innovation, and therefore were not at first produced by the Russians themselves, but were imported from Western Europe. Also, gunpowder and firearms did not immediately become familiar to the general population. Vodka, however, as the first real mass industrial product, necessarily had a greater impact on the country's economy, and its adoption by the masses caused a greater social shock.

Until this time all trades and all production had been traditional, based on skills, methods and tools developed by previous generations and already known for centuries, or at least for decades. Consequently, the significant number of people required to implement the monopoly of vodka production and sale had to learn something absolutely new. They had no teachers, traditions or skills, and were forced to break with their previous ideas. For the first time in that epoch all these people were engaged in service of an industrial rather than trade or handicraft character, in state production.

In general outline, these were the economic and social factors, the prerequisites, causes and consequences of the development of vodka production.

The large-scale sale of vodka took place through a network of state liquor establishments, the "Tsar's taverns". The transition to a new type of outlet, the ban on the sale of vodka in private shops of all types and especially in the inns, and also the appearance of the term *korchemstvo*, meaning illicit trade in vodka, all have potential for helping to establish the date when vodka appeared in Russia.

The Social, Moral and Ideological Consequences of the Appearance of Alcohol Distilling in Russia

Let us turn now to the social phenomena which may serve as a pointer to the advent of vodka in a society which had not known it previously.

One of the noticeable peculiarities of vodka, as a product and as a commodity, was the fact that it exercised a corrupting influence on the old, closed medieval society, permeated with ancient traditions. At a single blow vodka destroyed social, cultural, moral and ideological taboos. In this respect it acted like an atomic explosion in the stagnant calm of patriarchal feudalism. The effects of the appearance of vodka are particularly easy to detect in the social and cultural fields, details of which are preserved in the documents of the epoch, from legal edicts to literature. Hence the signs of the advent of vodka will not be statistical indicators pointing to a rise in drunkenness so much as the rise of new social relationships, increasingly free of the fetters of patriarchal morality; the rise of new social conflicts; and a noticeable acceleration in the social stratification and class differentiation of society.

Of special significance was the use of vodka by the ruling classes as a new weapon of social policy. The use of alcohol to demoralize and debauch the peoples of the North, something well known in the nineteenth and twentieth centuries, was foreshadowed by the habituation to alcoholism of the native Russian population of Moscow state.

The signs that vodka had emerged on the scene are thus readily detectable through an analysis not only of economic, but also of social and cultural materials from the end of the fourteenth to the beginning of the sixteenth century. Finally, the creation of new establishments for the storage and sale of vodka was also reflected to a significant degree in the social and ideological spheres.

Simply on its own, the presence of a new social category of tavern-keepers, people whose main quality must have been an extreme cynicism, inevitably caused the rise of conflicts never before seen in medieval society. The "ideology" of the tavern-keepers posed a direct challenge to official Christian morality. And since the source of the new social contradictions was none other than the Tsar's government, moral conflicts inevitably became social ones. At the same time as the Church was hurling thunderbolts against drunkenness and the insolence and godlessness that accompanied it, the tavern-keepers were receiving instructions "not to drive drunkards (*pitukhi*) away from the Tsar's taverns in any circumstances"; that is, they were being called on to increase the sale of vodka by all possible means. All this could not fail to find, and undoubtedly did find, a reflection in the most diverse manifestations of public opinion, in social conflicts, and in the appearance of a new type of "insolent" person, no longer bound by medieval morality and traditions.

Such people rapidly came to inhabit a new social layer: the lumpenproletariat, or to use the terminology of the time, the *posadskaia golytba* (urban poor) – those who were forever half-starved, spiteful, cruel, cynical, unfettered by any of the norms of the people in general, and whose energies could be directed into any channel for the price of a "bucket" of vodka.

Following the appearance of vodka the numbers of destitute people among the urban population rose sharply. The recruits to this social stratum came on the one hand from the families of ruined drunkards of all classes, and on the other from the less industrious part of the alcohol-ravaged *posadskaia golytba*. This jump in the numbers of the poor, a jump whose abruptness contrasts with the usual process of impoverishment arising out of natural economic factors – famine, class differentiation and so forth – testifies to the fact that at this time vodka played a role in creating poverty. This role was not only direct, but also indirect; for example, through a sharp increase in the number of urban fires, in which whole towns and villages were consumed. Finally, the incidence of disease, including epidemics, is also linked with the rise of alcohol distilling and with the widespread sale and consumption of vodka.

These in general outline are the social, cultural and moral factors, the consequences of the appearance of alcohol distilling or the pointers to it, which can be identified through a study of the history of the period from the end of the fourteenth to the beginning of the sixteenth centuries, and through which a date for the rise of vodka production can be established.

Where Was Vodka First Made?

Our study will include a detailed examination of the socio-economic, external and internal political, cultural and ideological developments of the epoch from the end of the fourteenth to the beginning of the sixteenth century, in order to reveal how and when historical conditions underwent sharp changes. But before we begin these investigations, or set out to clarify how the changes were linked with the appearance and spread of alcohol distilling, we need to establish on what territory, in which state, and in which society we need to study these phenomena and search for these signs.

Although for hundreds of years the whole territory which later became the European part of the USSR was commonly regarded as "Russia", during the fourteenth and fifteenth centuries this area was occupied by a number of large independent states, in any of which the rise of vodka might have occurred. What were these states?

If we leave out of account a number of small principalities which were vassal or semi-vassal dependencies of larger states, then during the fourteenth and fifteenth centuries the region was occupied by the Golden Horde, the Grand Principality of Moscow, the Grand Principality of Tver, the Principality of Ryazan, the Principality of Nizhegorod, the Republic of Novgorod, the Grand Principality of Lithuania, the Crimean Khanate of Girei, the Genoese colony of Cafta, and the Livonian Order (at that time part of the Teutonic Order). Of these ten states two, the Golden Horde and the Crimea, could not have been the birthplace of vodka for a whole series of weighty reasons,[2] and therefore they can be excluded immediately from the reckoning.

The western states of Lithuania and Livonia require closer examination. Lithuania was the main European centre of the trade in honey. Here the resources had not been exhausted to the same extent as in the Russian states. The wild honey gathering of the northern part of the Lithuanian state began to be supplemented in the fifteenth century by beekeeping in the south, which was later a part of Ukraine. At the end of the fourteenth century the production of alcoholic beverages in Lithuania remained firmly based on honey, the more so since the demand of other European countries for honey and mead, and thus their prices, had risen strongly. Besides, in 1386 Lithuania had become a Catholic country. A side-effect of this was that it could not allow any curtailment of its honey industry, but was obliged to expand it and broaden its productive base, developing beekeeping to supplement the ancient practice of raiding wild hives. This was because Lithuania was seen in Rome as the main source of beeswax for the European candle industry, which was centred in Catholic monasteries. The Roman Curia demanded that Lithuania develop beekeeping, and this was done to such an extent that the Lithuanian economy became disproportionately dependent on the production of honey and wax.

While abundant honey was available as a raw material for the

production of alcoholic drinks, Lithuania was poorly supplied
with grain. In Lithuania proper and Belorussia the principal
cereal was oats, and in second place was rye; the harvests of
these grains barely supplied the basic needs of the local population. In the southern, Ukrainian part of the Lithuanian state, in
Volynia and in the Kiev and Bratslav regions, natural conditions
for agriculture were superb. But constant wars and raids by
nomadic peoples (especially the Nogaitsy and the Crimean
Tatars), and also the extremely sparse population of these
regions, made developed and profitable grain cultivation impossible. The Ukrainian Cossacks who were settled here in the second
half of the fifteenth century to defend the border zone against
nomadic raiders and to keep watch on the region did not themselves engage in agriculture to any notable extent. In part they
employed the services of tenant-hirelings; that is, semi-slaves who
were forced to hand over a large part of their harvest to the
owner in return for the use of the land, and who therefore were
entirely uninterested in raising the efficiency of agricultural production.

Hence in the "flourishing land of the Ukraine", right up to the
second half of the seventeenth century when Ukraine was united
with Russia, the development of the economy on the basis of the
region's natural wealth could not go ahead. The Ukraine supplied its own wants, bountifully in fact, but on the basis of the
isolated farmstead. As an overall economic, and even more as a
state organism, the Ukraine was practically an empty space. The
idea of organizing there a state-run industry such as alcohol distilling, which would have involved finding reliable people to exercise control over every aspect from acquiring the raw materials
to selling the finished product, would have seemed fantastic in
the conditions of the Middle Ages.

In the 1540s Lithuania saw the creation of the Zaporozhe
Sech, a state possessing both military strength and exceptional
discipline in its internal life. But by this time alcohol distilling
was already well developed in, for example, the Moscow state, as
well as the neighbouring countries of Poland and Sweden.
Consequently, one can be certain that neither Lithuania in general nor any of its regions – Zhemaytia, Old Lithuania, the
Liflyanty (Latgalia), Belorussia, Black Russia, Volynia and the

Ukrainian regions of Kiev, Bratslav and Podolia – could have been the site of the first appearance of alcohol distilling in the fourteenth or fifteenth centuries. It is historically impossible.

In Livonia and Cafta, small states set respectively on the north-western and south-eastern edges of Eastern Europe, but possessing close historical, economic and cultural links with Western Europe, the rise of grain spirit production was impossible for economic reasons, above all the lack of raw materials. Technically and theoretically, not only the discovery of alcohol distilling but also the production of alcoholic spirit on an experimental scale were both fully possible. The Genoese, the colonial masters of Cafta, knew of distillation and the obtaining of alcoholic spirit from grape wine in the 1340s.

Until 1309 the Livonian knights, who had close links with the Teutonic Order, constantly visited the latter's estates in the Veneto and Transcarpathia. It is highly probable that they maintained contact with Italians throughout the fourteenth century and that they became familiar with the production of alcoholic spirit. In 1422 representatives of the Teutonic Order in Danzig demonstrated "burning wine" (*goriashchee vino*) – alcoholic spirit. This could have been spirit they had prepared themselves, or an import from Italy or France. Whatever the case, there is no difference of principle here: any production they carried on could only have amounted to the experimental preparation of small batches of alcoholic spirit (it is even more doubtful that this was grain spirit) in monastery laboratories. Before there could be any application of this discovery to the process of production (to use the terminology of our own day), a prolonged period would have to elapse, and still more time would be required before vodka was transformed from a laboratory sample, a chemical substance, into an important factor in the economy and society. It was only in this sense that vodka could really become vodka, in the full meaning of the term.

This emergence of vodka as a social concept also has a direct relationship with its nature as a commodity, to its reliably meeting high standards of quality. For it would not be until vodka became a significant social factor that society, and the state as the main agent of society, would begin to devote serious attention to vodka, and to its quality, composition, and conformity to a standard throughout the state. This is entirely understandable, since vodka, as a commodity of great importance to the state, whose production and sale the state reserved for itself and which became practically the equivalent of money, needed to reflect the high prestige of the state. In the Middle Ages any change in the quality of vodka was perceived as a crime against the state, like counterfeiting its currency.

Every government, therefore, has shown concern for the quality and uniformity of the spirit over which it exercises a monopoly. Officials who might offend against its rules are always individual, lowly placed, venal functionaries of local administrations, and their transgressions always harm the prestige of the government; these officials do not in any way act with the blessing of the ruling power. Moreover, such violations of the state standard of quality are always felt acutely by society and by the people, who in cases of gross shortfalls in the quality of goods subject to state monopolies react in an extremely sharp fashion, from riots to major uprisings.

The history of Western Europe and of Russia includes many examples of this: a "salt and copper riot" in Moscow, provoked by a rise in the price of salt which was, moreover, of low quality, and by the replacing of the standard silver coins with copper; a "silver riot" in Novgorod, caused by the replacing of the former "heavy" *grivna* with small minted coins; and analogous riots in various countries caused by reductions in the weight of silver money or changes in the standards, or the use of false weights.

During the period of the vodka monopoly in Russia there was not a single recorded instance of anything which could be described as a vodka riot, either under the old monopoly (from the fifteenth to the seventeenth century) or the new (from 1896 to 1917).[3] Discontent was aroused only by the excise system and by the "tax farm" (*otkup*) under which the state in effect hired out its monopoly, a situation which led to abuses in particular localities. This once again testifies to the fact that the concept of value for money, of the high quality of vodka as a special product imbued with social significance, is indissolubly linked with the social and legal position of the drink. As a result, when we speak of the rise of vodka production, we ought not to divorce this question from the status which vodka acquired after its appearance. If vodka arose as the result of an experiment, as a sample (even if this sample was prepared in a quantity of hundreds of litres), but did not become a trade commodity – and moreover, a commodity accorded special recognition by the state and enjoying privileges in the form of a monopoly – then it could not yet be called vodka in the full sense of the word.

Only the establishment of a definite legal and social status for vodka, recognized by the state, could ensure its quality as a beverage. This quality might not necessarily be very high (something which is readily understandable, since in any particular historical period it would depend on the general level of knowledge and technical development), but it would never be arbitrary and subject to fluctuations like those of *samogon*; that is, illicit vodka produced without regard for the state standard. Nor could this quality be low in relation to what could be achieved by the techniques of the time, since the state monopoly would always strive to achieve the highest standard of quality possible in a given period. This would be dictated both by a concern for the prestige of the state, and by a consideration far removed from altruism – a purely businesslike desire on the part of the first-ever state firm to ensure that the product on which it enjoyed a monopoly could not be faked, and would be recognized by any consumer on the basis of its quality.

Hence, vodka really became vodka only from the moment when it became a product cherished and protected by the state. In the absence of this we are dealing merely with an experiment in preparing grain spirit, and it is impossible to talk about the production of vodka. It is therefore quite obvious that casual accounts of the presence of grain spirit or of the display of distilled spirit on the territory of the Teutonic Order in the early fifteenth century, or other similar evidence, cannot serve as a basis for concluding that this was where vodka was first produced. This is even more apparent when we consider that the account relating to the Teutonic Order was not followed by any further evidence pointing to the establishment of grain alcohol production or of a vodka trade in Livonia. And indeed, such developments could not have occurred, simply because after its defeat near Grünwald in 1410 the Teutonic Order never recovered its strength, and passed out of existence in 1466 after spending most of the previous half century in constant wars.

In 1422, the date of the reference to the display in Danzig of "burning wine", the Order concluded the Peace of Melnesky with Poland, and it is possible that as part of the peace celebrations the Teutonic knights imported alcoholic spirit or prepared it in the Order's monastery laboratories to Provençal or Genoese recipes. But we have absolutely no knowledge of what type of spirit this was, of whether it was made from wine or grain. Even the name "burnt wine" (*Branntwein*), later applied to grain spirit and vodka in the Germanic world, tells us nothing in this case, primarily because it was used merely to describe an impression. The fact that this name was at some times and in some places applied exclusively to grain (or later, to potato) spirit means nothing in the present instance. Here the word is not a generic term, and consequently we do not have the right to place on it the interpretation we normally give to the terms "grain spirit" and "vodka".

In this connection it is important to recall that in the Germanic countries grain spirit and the term used to designate it appeared only after the Great Peasant War, that is, somewhere in the 1530s or 1540s. When Martin Luther translated the New Testament from Greek around 1520 and encountered there the Aramaic word *sikera*, in a context that made it clear that wine was not meant, he could not find a German equivalent for it and therefore described it merely as "strong drink". This shows that the phrase "burning wine" was merely an attempt to describe a liquid which could in fact be set alight, and not a forerunner of the later German name for spirits, which refers to their having been "burnt"; that is, distilled – the word for both is *gebrannt*. It was only considerably later that "burnt wine" became an accepted term. This occurred in Germany during the early to middle decades of the seventeenth century, during the Thirty Years War.

The Teutonic Order and Livonia could not have become the founders of vodka production because they did not possess the economic prerequisites for developing this industry. There were constant wars on the territory of the Order. The shortage of labour became chronic from the fifteenth century as a consequence of severe population losses; these resulted from the wars and from the massive flight of peasants into the neighbouring states of Poland and Novgorod Russia to escape the tyranny of the Order and heavy feudal labour obligations (in Poland and Novgorod serfdom did not exist at this time). Finally, it was necessary to import grain from Poland for bread, since local supplies were inadequate. All these factors combined with the general political and economic debility of this military-theocratic state to make the establishment of alcohol distilling here at this time completely impracticable. Therefore we must also exclude Livonia from our search for the homeland of vodka.

We can be certain, on the basis of clear sources, that alcoholic spirit was prepared in the Genoese colony of Cafta. However, we can be satisfied that this was not grain spirit. Although there is no definite indication of the starting material for distillation, we can rely on the fact that the Genoese possessed grape wine. Moreover, the Genoese had obtained their knowledge of distillation from the Provençals, who used grape wine.

In 1386 a Genoese legation, which had been sent to Lithuania by way of the Moscow principality when the Lithuanian ruler Vitovt and his people were converted to Catholicism, displayed samples of alcoholic spirit to the Moscow boyars. This was not, however, presented as a drink, but as a medicine. In 1426, while on their way to Lithuania for the Troki Congress, where Vitovt was to be confirmed as king, the Genoese displayed further samples of *aqua vitae* to the court of Vasily III; but the preparation was considered excessively strong and quite unfit for drinking. The Genoese legation also took *aqua vitae* to Troki, for Vitovt, but the description of the seven-week festivities contains no indication that this gift was perceived as a drink, or that it was offered to guests as one of the alcoholic beverages associated with the feasting. It seems likely that in this case too it was considered solely as a medicine.[4]

It is interesting and suggestive that the displays by the Teutonic-Livonian knights from Danzig and the Genoese from Cafta of this Western European innovation, wine spirit or *aqua vitae*, took place at almost the same time, in 1422 and 1426. In both cases an opportunity had arisen to acquaint Eastern European courts – Russian, Lithuanian, and possibly Polish – with the achievements of Western science and technology, with the aim of demonstrating the power and skills of the West.

However, at this stage the matter did not go beyond experiment and demonstration. Alcoholic spirit made no impression on the "eastern barbarians" either as a medicine or as a trade commodity. Neither one side nor the other thought or speculated about the possibility of producing grain spirit.

Hence, neither Livonia nor Cafta served as the centre for the rise of alcohol distilling or of vodka. Cafta was completely destroyed by Tamerlane in 1395, and its lands were taken over by the Crimean Khanate of Girei; it ceased to exist not only as a state and an economic formation, but as a cultural and artisanal centre in general.

We shall now examine the situation in the Russian states. The oldest and most powerful of them, the republic of Novgorod, was oriented economically, and to some extent in its external political relations, towards the Baltic and the coastal regions of the North. Novgorod conducted an extensive trade with the Hanseatic cities and with Denmark, Sweden, Holland, the Teutonic Order, and to a lesser degree Poland and Lithuania. The main items exported by the Novgorodians were furs, flax, hides and hemp, and there were also exports of wax and silver. From the principality of Moscow the Novgorodians bought grain. For many years Novgorod had exchanged its wares for grape wines from Western Europe, including Rhine wines and burgundy. The import of grape wines had been a tradition in Novgorod since the tenth and eleventh centuries; in addition, Novgorod re-exported a proportion of these wines to the principalities of Moscow and Tver, in exchange purchasing honey from Moscow for export to Western Europe.

Therefore, despite the economic might of Novgorod, the conditions did not exist there for the rise of alcohol distilling. There was no surplus of grain; the republic did not possess its own developed cereal agriculture. There was no shortage of imported grape wine, and there was a tradition of consuming it. Moreover, from the twelfth century a strong tradition of beer-brewing had grown up in Novgorod, mainly of a ritual character, and taking the form of communal production. Beer was brewed by the whole street or neighbourhood, with all households sharing the costs.

Numerous careful archaeological excavations carried out over many years in Novgorod, Pskov and other towns on the territory of Novgorod Russia have also failed to provide any material evidence that the production of grain spirit arose in this region. Thus there is neither documentary evidence (in birch-bark texts), nor concrete archaeological evidence, nor any economic or historical basis for considering it possible that vodka originated in Novgorod.

Moreover, when alcohol distilling was already developed in the Moscow state it did not take root in Novgorod, and vodka was imported from Moscow. Consequently, the Novgorod republic fails to qualify as a territory in which vodka production might have arisen.

The next state bordering on Novgorod was the Grand Principality of Tver. During the fourteenth century this state underwent extremely difficult times. Its territory progressively shrank as it was hemmed in by the principality of Moscow, which subjected it to constant military and diplomatic attacks. Moscow by the middle of the fourteenth century had spread its influence far to the north, taking possession of Vologda, stretching its control beyond the Volga, and was already reaching out towards the approaches to the Urals; but it ran into the border posts and customs barriers of the Tver principality only eighty kilometres from the Kremlin, at Volokolamsk. Naturally, the princes of Moscow did everything in their power to annihilate Tver as a state. Eventually they terrorized the inhabitants of Tver by blockading the state's borders, isolating it from needed imports, and harassing the peasants with raids and the destruction of their crops so that in the end they began to flee into the Moscow state, where they were shown "favour". The prince of Tver was thus deprived of the support of his subjects, and was forced to abandon the city and flee to Lithuania.

It is absolutely clear that the struggle by Moscow to wear down the resistance of Tver, a struggle which had continued for more than a century, did not allow the inhabitants of Tver to concern themselves with anything but defence. Conditions in Tver, naturally, did not favour the rise of alcohol distilling, the more so since the city was having enormous difficulty in supplying itself with grain in the face of a constant blockade from Moscow.

It remains to be explained how the position stood with the Russian states lying to the south-east of Moscow – the Grand Principality of Ryazan and the principality of Nizhny Novgorod. Before the battle of Kulikovo in 1380 there was no possibility of alcohol distilling arising in Ryazan, any more than in the other Russian states. After Kulikovo the position of Ryazan, the only one among the Russian states to have supported Khan Mamay and to share in his fate, was extremely difficult. Ryazan was unable to recover either politically or economically after the battle. Finding itself in perennial alliance with the Horde and with the Crimea, and with its relations with Moscow strained and hostile, the Ryazan principality in the last century of its independent existence became effectively transformed into a nest of intrigues and a place of sanctuary for fugitives from various states – Tatars, Nogays, Lithuanians and Russians. Within the frontiers of the principality there also lived members of the Moksha, Erzya and Chuvash peoples, something which increased still further the national diversity and instability of this transitional border state.

Accordingly, the economic situation of the Ryazan principality was extremely unstable. The Russian inhabitants, who made up only about half the "Ryazan people", bore the main weight of feudal obligations and were the main producers of grain. The others did not engage in agriculture. As a result it was only in good years that Ryazan enjoyed a surplus of grain, which went for export. The rest of the time Ryazan barely supplied its own grain requirements. The possibilities for expanding the sown area were limited, since a substantial part of the principality was made up of the boggy Meshchera district, and of forests which comprised the honey-gathering and hunting territories of the Mordva people and of the allied Kasim Tatars. Also limiting the area sown were the fears held by the Russian population of pillage and devastation in Ryazan's condition of political instability.

Finally, the development on the territories of Ryazan of ritual mead making and beer-brewing, under the influence of Mordva national traditions, formed, as elsewhere, a hindrance to the rise of alcohol distilling. The preparation of a honey beer, *pure*, was of its very essence a natural obstacle to the development of grain alcohol production, since it was based on fundamentally different techniques, and tended to preclude any other thinking on how to produce alcoholic drinks.[5] Such situations have recurred widely; in Bohemia, for example, where classic beer-brewing had been carried on since ancient times, the production of grain spirit arose only very late, in the nineteenth century, and then under the influence not of national but of external factors. The same occurred in Ryazan.

The principality of Nizhny Novgorod was in a favourable situation compared with Ryazan, not being subject to the same destabilizing factors. It benefited from a long-term alliance and close economic relations with Moscow and, despite its closeness to the Tatar lands, was not devastated in the way Ryazan had been. Its escape from this fate was due in part to impenetrable forests which covered a large part of the territory. However, the development of agriculture here was rather weak. The economy was based on fishing, on the hunting of wild game and of beavers, on honey-gathering and on the making of bast shoes, baskets and mats. Stock-raising, mainly of sheep, was developed to some degree; so too were wool-washing and felt production. All this made Nizhny Novgorod a hunting and handicraft reserve of Moscow, a natural trading partner with interests in selling its forest products to the population of the Moscow state. But this economic orientation hindered the development of grain cultivation and of agriculture in general; also, if agriculture were to be carried on the virgin forest had to be cleared, which was an extremely laborious process. Grain here was valued highly, surpluses never arose, and in years of famine grain was imported from Moscow.

The presence of significant stocks of honey in proportion to the population in Nizhny Novgorod helped maintain the ancient tradition of mead production, as in neighbouring Ryazan. There was therefore no need to find a replacement for the old alcoholic drinks.

Taken together, all these circumstances provide grounds for concluding that alcohol distilling could not have arisen on the territories of Nizhny Novgorod.

Consequently, none of the independent Russian states – Novgorod, Tver, Ryazan, or Nizhny Novgorod – can be seen as a potential birthplace for the distillation of grain spirit in the fourteenth and fifteenth centuries. These states all lacked one or another of the preconditions essential for the emergence of vodka.

It remains to be considered how matters stood in the Grand Principality of Moscow. By the middle of the fourteenth century the Moscow state was already the largest and most powerful of the Russian principalities. In the first place, it was the most densely settled, since an extremely diverse population flowed into Moscow during the thirteenth and fourteenth centuries, seeking protection from the depredations of the Tatars. Second, the degree of development of agriculture was much greater in Moscow state than in the neighbouring principalities. Third, the Moscow principality possessed many more towns and a much larger urban population than the neighbouring states. This stimulated the development of agriculture, since peasants could easily sell surpluses of grain and other agricultural products to the townsfolk. The prices for agrarian produce, especially in Moscow itself, were the highest in Russia, since the non-productive population was constantly growing. Fourth, the principality of Moscow was the site of the largest and best organized monasteries, which were improving their equipment and methods, introducing new trades and influencing the development of money transactions. The monasteries produced a surplus of goods, and money was important to them for obtaining expensive church utensils and icons, for foreign trade, and also for building churches and defensive fortifications and for obtaining weapons. Fifth, the demand for honey and wax was much greater in Moscow, while as a result of the cutting down of forests the gathering of wild honey had dwindled almost to nothing. Beekeeping was as yet unknown. Honey and wax had to be imported; they were either bought, or extracted from distant colonies as a tribute. Sixth, grain was intentionally being produced in quantities greater than local requirements, with a view to exporting it to foreign markets – Novgorod and the Horde. Seventh, an urban population which was accustomed to spells of idleness (especially building workers and seasonal workers in Moscow), needed to be distracted from its cares. The problem of finding cheap alcoholic drinks for mass consumption had already arisen in the middle years of the fourteenth century in connection with a series of Moscow riots, primarily of building workers engaged in the construction of the Kremlin (their number reached as many as 2,000 to 2,500). Eighth, Moscow as a political centre was linked with Western Europe, Byzantium, Sweden and Denmark. Here were ambassadors from India, Persia and Turkey, while travellers came from Greece, Italy, Bulgaria, Moldavia, Wallachia, France, Germany, Sweden, Holland, Poland, Denmark, Schleswig-Holstein, Livonia, Estonia, Finland, Genoa, Venice, Serbia, Hungary and Bohemia. Moscow therefore enjoyed much greater possibilities for borrowings, for cultural collaboration, and for the penetration of new influences and ideas. This remains true even if we consider how closed the society of that time was, and the difficulty of contact with foreigners even at the court.

Moscow therefore had greater potential than the other states as a base for the rise of any new productive trade, and for any type of technical innovation including the production of grain spirit. This is why we must conduct our analysis of the setting needed for the rise of alcohol distilling and vodka production first of all on the basis of a study of the history of the Moscow state. It was in Moscow that a centralized state first arose in Russia; it was here, therefore, that the political, economic, social and technical preconditions for vodka production were concentrated.

It should be noted that several Russian historians who have studied the legal institutions of the Moscow state have stressed that such a product as grain alcohol, requiring the introduction of a state monopoly, could arise only under the conditions of a centralized, autocratic state. This in itself was held to prove, at least indirectly, that vodka arose in the Moscow principality and not in the neighbouring Russian states. It is significant that already in the seventeenth century, and then throughout the eighteenth and nineteenth centuries, the expression "Moscow vodka" was firmly implanted in popular speech. In the twentieth century it became the official name of one of the brands of vodka, and so is no longer perceived as it was in the past. But in previous centuries it had the same ring to it as the names of other products typical of a particular city or locality: Moscow loaves, Tula or Vyazma buns, Valdai bells and bread rings, and Vyborg bread twists. All these names are intimately linked with the fact that a particular product originated in a certain place, and that this place continued to specialize in the production of the commodity. It is important to note that the continuation of this specialization brought with it an effort to safeguard the reputation of the product, and so obliged the makers to improve its quality in any way possible. After a few decades the names "Moscow vodka" and "Vyazma buns" were perceived as honorific, as guarantees of quality.

It is significant that even considerably later, in the epoch of the development of capitalism, when vodka had begun to be produced in many parts of Russia, it did not come to be called Kiev, or Vinnitsa, or St Petersburg, or Ryazan vodka – it remained Moscow vodka. This testifies once again to the early origin of this name, and to the fact that such names cannot arise in later times; or if they do, they acquire neither broad popular currency, nor the status of common nouns.

This is a fact of historical significance, which should be given particular emphasis; in the light of it, any idea that vodka could have originated in some other part of Russia appears quite fantastic. For example, a suggestion that vodka arose in 1474 in Vyatka, then a colony of Novgorod and later of Moscow, is quite untenable. No one has ever heard of Vyatka vodka; yet if vodka had been developed there, this would soon have become embedded firmly in the popular consciousness, just like Moscow vodka. The name "Moscow vodka" became attached so firmly to the drink because the fact of its production there was reinforced by its sale, which in the first years of the existence of vodka took place only in the city of Moscow, where the first "Tsar's taverns" were established. Consequently, the term reflects the historical fact that vodka arose originally, and moreover exclusively, in Moscow, and testifies to the uniqueness of vodka as a specific Moscow product, at first unknown elsewhere. This obliges us to look carefully at events in the history of the Moscow state during the fourteenth and fifteenth centuries, in order to determine at what time in this period grain spirit production could have emerged.

The Political and Military History of the Moscow State during the Fourteenth and Fifteenth Centuries

Before we turn to an analysis of the historical situation, we should collate all the available chronological evidence of the rise of alcohol distilling or of vodka in Russia, even if this evidence appears misleading or insufficient. In the present case it is important to be able to take in at a glance the information which our predecessors had available to them when they considered this question, and which is already known to us without the need for additional research. It is also important to be able to compare the Russian case with the chronology of the development of distilled spirit in other European countries. This will define more clearly the place occupied by Russia, and will narrow the ambit of our search.

Early 12th century: first European mention of distillation of alcohol, in *Mappa clavicula* – probably inspired by the work of Arab scientists, who had in turn learnt the technique from Hellenistic Egypt.

1334: preparation of alcoholic spirit by Arnold Villeneuve in Provence.

1360s: preparation in a number of Italian and southern French monasteries of a highly concentrated alcoholic spirit named *aqua vitae*.

1386: Genoese legation in Moscow displays *aqua vitae* to Moscow boyars and foreign apothecaries.

1422: knights of the Teutonic Order display *aqua vitae* in Danzig.

1426: Genoese legation on its way to Lithuania visits Vasily III and presents him with *aqua vitae* as a medicine.

1446: one of the last mentions of mead in Western sources in the Old Slavonic language, which suggests that from roughly the middle of the fifteenth century, the large-scale export of Russian honey to Central and Western Europe ceased.

1474: *korchma* (strong drink) is mentioned in the Pskov chronicle ("do not bring any *korchma* ...").

1485: the first English gin is prepared at the court of Henry VII.

1493: at the time of the conquest of the Arsk land (Udmurtiya), the preparation of *kumyshka* (milk vodka) is reported.

1506: a Swedish source relates that Swedes had brought from Moscow a special drink called "burnt wine".

If we ignore the first two of these dates, since any discovery made by alchemists remained a secret for many decades,[6] all our information about alcoholic spirit in Europe and about its production stems from the period between 1386 and 1506. The

latter date, we may consider, marks the point at which the production of vodka expanded beyond the bounds of narrow
national control and became a general European phenomenon,
crossing state boundaries. It is on this period of 120 years, therefore, that our search must be concentrated.

Perhaps we should begin our study earlier than 1386; say,
from the middle of the fourteenth century? Otherwise it might
seem that by beginning our search at the moment when the
Genoese brought *aqua vitae*, we are linking the rise of alcohol distilling with a Western European discovery. This, however, is not
the point. For a start, the surprise aroused by the demonstration
of *aqua vitae* indicates that such a product was unknown in
Moscow at that time, a more than sufficient reason for taking
this date as a marker. There is also another indirect proof of the
fact that grain spirit and distilling were unknown in Moscow at
least until 1377. This is a detailed description of a tragedy on the
river Pyana (the name means "drunk") in which a drunken force
of the home guards of the princes of Pereyaslavl, Suzdal,
Yaroslavl, Yurev, Murom and Nizhny Novgorod was almost completely wiped out by a small force under the Tatar Tsarevich
Arapsha on 2 August 1377. Moreover, Prince Ivan Dmitrievich
of Nizhny Novgorod drowned together with his entire inebriated
retinue. In the chronicle, which provides scrupulous details of
their intoxication, the only drinks mentioned are mead, ale, beer,
and Mordva *pure*, which the warriors of the prince of Suzdal
drank while waiting in the local villages for the battle to begin.[7]

We thus have a precise pointer to the domestic, patriarchal
character of the production of alcoholic drinks at this time, and
to the corresponding nature of the drinks themselves. So if we
wanted to be even more precise about the period in which the
origins of alcohol distilling in Russia are to be found, we could
conduct our search from 1377 instead of from 1386.

The only point which arouses suspicion here is the fact that,
just as in the case of the great bout of drunkenness in Moscow
five years later, and the invasion by Tokhtamysh in 1382, the
character of the drunkenness induced by domestically produced
drinks was very different from that which was recorded, let us
say, fifty years before. It was much heavier, having a paralysing
effect on people. This could be explained by the use of alcoholic

spirit of very inferior varieties; but it is more likely to mark the beginnings of pot distilling, which gave a strong but grossly impure and poisonous product – obtained, however, from a honey base.

It is interesting to note the significance of the name Pyana (or Pyanaya) river. There is no doubt that the origins of this name lie in historical events. The name is not recorded until after the disaster of 1377. The saying "Beyond the Pyana the people are drunk (*piany*)"[8] dates from the same time. Such a saying could only signify that on the territory beyond this border river, in the Mordva lands, the population brewed a different and more intoxicating "herb wine" than the Russians of Vladimir, Suzdal, Nizhny Novgorod and Moscow were used to. The name of the river, of course, was thought up later by Russians. It might originally have borne some similar-sounding Finnic name, such as *Kaiana* or *Kaiani*.[9] Whatever the case, the mass drunkenness and resulting total defeat of the Russian forces so struck people's imagination at the time that it was recorded both officially in the chronicle, and in folklore.

Having established this well-known event as a starting-point in our search for the date of origin of alcohol distilling in Russia, we can begin to examine events between 1377 and 1506 – a period of 130 years. During this time great external political events affected the Moscow state.

Fourteenth century

1371 and 1375: conquest by Novgorod of Kostroma and Yaroslavl. Plunder by the Novgorodians on the Kama river, although Moscow is already beginning to lay claim to these territories.

1375: Moscow is victorious in a bloody war, but Tver forces it to give up plans to seek the status of a Grand Principality.

1376–9: Novgorod is still sufficiently strong to sell Muscovite prisoners to the Volga Bulgars and Tatars.

1377: the forces of the small independent principalities of Vladimir and Suzdal, allied with Moscow, suffer a crushing defeat on the river Pyana at the hands of the Tsarevich Arapsha.

1380: battle of Kulikovo. The forces of Khan Mamay are defeated by the united strength of the Russian principalities (apart from Ryazan and Novgorod).

1382: capture and sack of Moscow by Khan Tokhtamysh, following mass drunkenness of the townsfolk and army. Flight of Dmitry Donskoy to Kostroma.

1385: Novgorodians seize the property of the Grand Prince of Moscow in retaliation for repressions carried out by the Moscow forces to avenge the sale of Muscovites to the Tatars as slaves.

1386: twenty-six provinces and districts, including some subject to Novgorodians, unite under the leadership of Moscow and lay siege to Novgorod, levying a tribute of 8,000 rubles in silver ingots. Novgorod pays the sum.

1388: the son of Dmitry Donskoy escapes from captivity under the Horde. The Tatars continue to hold the princes of Nizhny Novgorod and Suzdal as hostages.

1389: war between Moscow and Serpukhov, and between Moscow and Ryazan.

1389: death of Dmitry Donskoy. Establishment of rights of majority in the succession to the Moscow princedom. The election of grand princes is abolished. The heir to the throne is no longer the oldest among the prince's kinsmen, or the brother of the prince, or an uncle of the prince's son; but the son himself, however young.

1391: first punitive expedition mounted by Moscow against colonies of Novgorod, involving attempts to seize them and exact tribute.

1391: treaty between Novgorod and the Hanseatic League, Livonia, and Estonia (Reval) against Moscow. The Novgorodians refuse to break their peace with the Germans at Moscow's demand.

1392: Moscow annexes the princedoms of Suzdal and Nizhny Novgorod.

1395: Tamerlane reaches Yelets and Ryazan, razes Cafta, conquers Transcaucasia, and destroys the capital of the Golden Horde.

1395: Lithuania seizes the principality of Smolensk. The border between Lithuania-Poland and Moscow extends along the line Borovsk–Mozhaysk–Ruza–Rzhev. The provinces of Kaluga and Tula are in the hands of the Poles and Lithuanians.

1397: attempt by Moscow to seize the Dvina territories and Zavoloche from Novgorod.

1398: Novgorod punishes the population of the Dvina territories, but surrenders Zavoloche.

1399: Lithuanians routed in a battle against Khan Edigey. Vitovt loses the Tula, Orlov and Kaluga territories, and the Tatars penetrate as far as Lutsk. Moscow moves its border westward to the line Smolensk–Kalug–Medyn.

1399: death of the Grand Prince Mikhail the Great of Tver, and conflicts within the Tver state. Moscow attempts to take advantage of the weakened condition of the vassals of Tver, the princes of Lamsk and Kashin, in order to annex their territories.

Fifteenth century

1404: campaign by the Moscow forces against the Novgorod colony of the Dvina lands; a pirate expedition of 250 ships is set up to plunder the Kama territories with the aim of obtaining silver.

1408: complete cessation, for the first time, of the provision by Moscow of tribute and gifts to the Golden Horde.

1412: resumption of payments of tribute to the Horde (until 1425) as a result of a repressive expedition mounted against Moscow by Edigey and the complete devastation of ten towns in the Vladimir and Moscow territories. A tribute of 3,000 rubles is levied, which the Muscovites pay gladly to prevent Edigey from taking Moscow.

1431: Vitovt imposes on Novgorod a tribute of 55 *pudy* of silver ingots; the Novgorodian *pud* or *berkovets* was equal to 16 kilograms, so this was the best part of a tonne of high-purity silver. Novgorod pays this sum over five months. The sale to Moscow by Novgorod of foreign (burgundy and Rhine) wines comes to an end.

1453: conquest of Byzantium by the Turks. Flight into Russia of Byzantine clergy, soldiers, government officials and artisans.

1458–9: conquest of the Vyatka territories, and the end of the Vyatka republic.

1460: complete and final end of tribute payments by Moscow to the Horde. Disintegration of the Horde into a series of small warring khanates.

1460–65: Cafta is seized and laid waste by the Crimean khanate.

Complete cessation of Moscow's trade with Italy (Venice, Genoa, Pisa). Imports of French, Italian, Greek and Spanish grape wines come to an end.

1466–72: journey of Afanasy Nikitin to India with the aim of finding an outlet to the East to compensate for the end of trade with the Mediterranean. Unsuccessful attempts to find a source of grape wines in India and Persia. Nikitin succeeds only in obtaining agreements on trade in fabrics, precious stones and spices.

1472: Great Perm is annexed to Moscow.

1475: beginnings of trade between Moscow and Persia.

1478: annexation of Karelia.

1478: Novgorod loses its independence and is annexed to Moscow.

1480: complete formal liberation of Moscow from all forms of dependence on the Tatars. Moscow begins to apply pressure on the khanate of Kazan.

1483: initiation of trade with Turkey and the Crimea in fabrics and fruits.

1483: conquest of the Yugra-Vogul principalities.

1483: conclusion of the first trade agreement with India.

1484: founding of the port of Arkhangelsk, and the beginning of trade with Holland.

1485: the principality of Tver comes to an end. Tver is annexed to Moscow by conquest. The princes of Tver are driven into exile in Lithuania.

What is the significance of these facts, which appear to be isolated and unconnected?

Most importantly, they allow us to see the sharp contrast between the positions of the Moscow state in the fourteenth and fifteenth centuries. Right to the end of the fourteenth century the military situation and foreign relations of the state were extremely unstable. At particular moments one even has the impression of a complete loss of the gains made in previous decades. Of course, the general historical trend was toward strengthening the position of Moscow. But while we can say this now, with the benefit of hindsight, it was not in the least obvious to people at that time.

The breakthrough came only at the very end of the fourteenth century. The defeat of the Volga Bulgars in 1399 opened up a whole period of consistent growth in the territory of the Moscow state, and consequently in its population and economic, military and political status; Moscow was transformed into a great power, an empire. All this accelerated the process of centralization. The fifteenth century was on the whole a time of advance, but this was not uniform. Three stages stand out.

The first of these stages lasted from 1399 to 1453. During these years the growing power of Moscow still encountered opposition from rival states, though with hindsight it is obvious that the days of these states were numbered. It was unclear to contemporaries how much longer this period of alternation between seemingly decisive victories for Moscow and renewed resistance from its opponents would last. And this, naturally, forced the Muscovites to harness all their energies to achieve their main goal: freeing themselves from external subjugation.

The second stage, from 1453 to 1472, though relatively brief, was characterized for the Moscow state by the advent of a completely new international situation. This period saw the culmination of the centuries-old quest for full political independence. External threats vanished entirely. Moreover, a new historic mission opened up: to become the successor of Byzantium throughout the whole of Eastern Europe and the Near East. Simultaneously with these enticing political perspectives an entirely new but not altogether agreeable situation developed in the area of foreign trade. The old external markets disappeared abruptly, and imports of grape wine, essential for the ritual needs of the Church, were completely cut off. The task of developing new external trade relations, and above all of finding a source of wine, became pressing. This challenge was met comparatively rapidly, especially given the conditions of the time.

The third period, also comparatively brief, was the most straightforward in its policy directions and historical outcome. In the years between 1472 and 1489 the Moscow state performed a dramatic leap to the east. In the course of these two decades the whole territory of the Russian state, as we are accustomed to regard it, came under Moscow's sway. To digest such vast acquisitions would be difficult for any state, but Moscow coped with the task. This demonstrates that the economic, political and social preconditions for these developments had all ripened; the changes that took place were not accidental, but historically inevitable. The main result of this process that is of interest to us from the point of view of our topic is the vigorous economic growth of the Moscow state through the establishment of a large market and through the drawing into the market economy of new territories and social groups. Such expansion can either lead to the concentration of capital needed for the rise of new forms of production, or it can itself be the result of the earlier establishment of the economic base required for expansion to occur.

In other words, either the prosperity of this period gave rise to vodka production, or vodka production brought prosperity. This suggests that alcohol distilling arose either at some point between 1460 and 1470, or in the period between 1472 and the end of the fifteenth century. One would hardly expect production to get under way in the troubled fourteenth century, even though various historians have not categorically rejected this possibility.[10] During the fourteenth century there was simply no period of peace long enough for the development of new forms of production, much as these might have been needed to build up stocks depleted by war.

We therefore need to look attentively at the economic history of the fifteenth century, comparing both the periods noted above, and only then deciding which is more historically probable as a time for the rise of vodka.

Economic and Social Conditions in the Moscow State at the End of the Fourteenth and during the Fifteenth Century

Below we shall set out the most important economic events of the period under examination, just as we did earlier in relation to the military and political events which characterized the general historical position of the Moscow state during the fourteenth and fifteenth centuries.

Fourteenth century
1367: construction is begun of the first brick Kremlin, in place of the wooden one, which was frequently burnt and required

constant rebuilding. The new construction work, extending over three decades, demands huge investments in materials; means of transport for shifting these materials along the Moscow river; payments to the architects, artists and tradesmen invited from Italy; payments to the best tradesmen from other Russian states, above all Novgorod, Pskov and Tver; the concentration in Moscow of a large army of building workers, more than 2,000 people, and of seasonal workers – the first "proletarians" in the Muscovite feudal state.

1370–90: erection of large fortified monasteries in Moscow and in the Moscow state – the Chudov, Andronnikov and Simonov monasteries in Moscow and its environs, and the Vysotsky monastery in Serpukhov. The costs of this construction are borne by the Church.

1372–87: construction of new towns with fortresses (masonry kremlins): Kurmysh (1372), Serpukhov (1374), Yamburg (1384), Porkhov (1387).

1375: Marten pelts are mentioned for the last time in an external political agreement of the Moscow state. From this time they are no longer accepted as payment for debts to the state.

1380s–90s: rapid growth in the role played by money in trade and in the policies of the Moscow state.

1384–6: Moscow demands and obtains from Novgorod the payment of a contribution covering a large part of the tribute paid by Moscow to the Golden Horde. This is represented as compensation for the services rendered by Moscow to the other Russian states, and principally to Novgorod, in protecting them from the Tatars.

1387–9: introduction of the silver coinage of the Grand Prince of Moscow and All Russia, with his portrait and crest. The acceptance of marten pelts and other furs as a means of payment in internal trade is ended. The first monetary reform in the history of Moscow takes place, involving the collection and burning of marten pelts, which are exchanged for silver.

1389: invitations are issued to foreigners – Genoese, Pisans and Greeks – to take up high Moscow government posts that demand special education. Foreigners serve as envoys, ambassadors, treasurers, tax collectors, administrators of provinces, and governors of distant colonies.

1389: introduction of firearms to the Muscovite state.

1393: establishment in Moscow of the first Russian factory for the production of gunpowder, employing foreign artisans.

1397–8: dramatic heightening of the struggle between Moscow and Novgorod for markets and for the possession of colonies.

Fifteenth century

1410: the government of Novgorod abolishes leather money in the Dvina territories and in Novgorod itself, making the transition to conducting its internal trade and trade with its colonies in Swedish and Lithuanian coins – *ortugi* and *groshi*. In trade with the West the former medium of exchange, silver ingots (*grivny*), remains in use.

1412: the Horde demands increased tribute payments and obtains these (until 1425).

1418: rebellion by debtors against creditors in Novgorod. Riot by Novgorod lumpenproletariat, demanding the handing out of public monies.

1419–22: four successive years of poor harvests in the Novgorod territories. Impossibility of harvesting grain owing to early snow (15 September). Widespread starvation.

1420: Novgorod introduces its own "national" silver money, similar to that of Moscow.

1446: contribution of about 20 million gold rubles (over 170 tonnes of gold) paid to Khan Mahmet by Vasily III ("the Dark"). This provides an idea of the means at the disposal of this spendthrift Grand Prince, whose treasury was, moreover, twice captured and ransacked.

1456: Moscow threatens to cut off supplies of grain to Novgorod, forcing the Novgorod assembly to accept that it cannot make laws without the sanction of the Grand Prince of Moscow. The regular levy imposed by Moscow on Novgorod becomes a permanent tribute obligation, no longer camouflaged as a contribution to the cost of defence against the Horde.

1460: complete cessation of tribute and other payments by Moscow to the Golden Horde.

1462: "Silver riot" in Novgorod; a refusal by the population to accept the new silver coins, the thin *cheshuiki*, which the Novgorodian government began to issue in an attempt to escape from a "silver famine".

1478: complete prohibition in the centralized Russian state of
trade by foreigners in imported goods on the internal market,
and also the decision not to use foreign merchants as commercial
agents of the Russian state in foreign markets (until 1553).

From these facts it can be seen that the political turning-point
which is apparent at the end of the fourteenth century becomes
even more distinct in the light of economic history. The period
between the late 1380s and the beginning of the 1410s, when a
revolution was occurring in the area of trade and commodity-
money relations, involving the need to replace old monetary
units and to introduce new means of payment for commodities
and labour, appears as an obvious boundary between the old and
new epochs. Our study of external politics suggests that until the
middle of the fifteenth century the trend of events was not yet
fully clear, and that the victory for the new type of relations that
was already implicit in external politics was not yet decisively
expressed. The economic facts, on the other hand, show that by
the period between 1380 and 1440 the scales were oscillating
between old and new; and that after the 1440s there could be no
doubt that the trend of development lay in the direction of victory for Moscow in the economic, military and political spheres.

If we note in addition that during the fifty years from the
1390s the population of the Moscow state doubled, and that
from the 1410s or 1420s overpopulation during poor harvest
years brought forward the transition to a new, more progressive
system of agriculture – the three-field system – then it becomes
clear that Moscow's economic victory was complete by the end of
the 1430s or the beginning of the 1440s. The question of who
would rule Russia had been decided, and the economic basis for
the creation of a centralized state was in place.

In no more than six to twelve years the regular rotation of
crops had brought about a sharp increase in grain production on
the same areas and with the same labour force. This allowed the
Moscow state to overcome with relative ease the shortages of
grain which existed in poor harvest years throughout the rest of
Russia, and which had ended disastrously for Novgorod. Moscow
was even able to accumulate surpluses of grain despite the

growth of the population, which was increasing not only as a result of natural increment, but to an even greater extent because of the stream of immigrants. By the mid-1440s the grain trade had come to involve the monasteries and the free peasants, who were selling grain to merchants for resale at higher prices in other principalities and outside Russia. These grain surpluses put the Moscow principality in a position that was unique at the time.

The transition to the three-field system was carried out in Moscow almost a hundred years earlier than, for example, in Sweden, Poland, Lithuania and the other Russian princedoms. That fact, which to this day has not been given due weight by historians, points clearly to the causes of the earlier establishment in the Moscow state than in the other states of Eastern Europe of the conditions for the rise of alcohol distilling. The early appearance of the three-field system in the Moscow state not only explains the especially healthy economic development of Moscow compared with the other Russian states, but to a significant degree alters the image of this state as historically backward compared with others in Europe. At least in comparison with its closest western neighbours – Denmark, Sweden, Lithuania, Poland, the Novgorod republic, the Teutonic Order and Livonia – the Muscovite state was more advanced as regards the organization and development of agriculture and the general level of development of the forces of production. Moscow outstripped Sweden,[11] for example, by some 80 to 100 years in these respects. In order to grasp the full extent of the advances which the rapid transition to the three-field system brought during the first fifteen to twenty-five years, including the scale of the surpluses of grain that were created in this way, we shall examine the example of Sweden, where the same process occurred a century later and where statistical data relating to it have been preserved. The good effects of the transition to the new system of agriculture lasted for no more than half a century, and a decline began to be felt after three or four decades. After this the expanded use of grain for alcohol production led to an abrupt worsening of the food supply.

The general picture of these developments is clearly visible from the example of Sweden. At least in its general features, the economic development curve in the Moscow state between 1446 and 1583 approximately matched the curve of development in Sweden a hundred years later, in the years from 1555 to 1661. The economic boom due to the transition to the three-field system brought about an increase in grain supplies of 200 to 250 per cent; in supplies of meat, of 300 to 400 per cent; and of fish, from 350 to 400 per cent compared with "normal" levels.

Annual per capita consumption of foodstuffs (kg)

| In Sweden: | 1555 | 1638 | 1653 | 1661 |
In Russia:	1446	1533	1555	1583
Grain	608	278	368	366
Meat	104.5	24	25.5	35.5
Fish	145.5	46	34	34
Salt	11.9	17	19	11

This is why the second half of the fifteenth century and the first half of the sixteenth century were among the most plentiful periods in Russian history. It is no accident that it was precisely in this span that the classical repertoire of the Russian national cuisine came into existence and that the first Russian cookery book was compiled, in 1537.

The history of the development of the Moscow economy during the fourteenth and fifteenth centuries allows us to conclude that alcohol distilling most probably arose in Russia at the apogee of development of agriculture, with its sharply increased surpluses of grain following the introduction of the three-field system. Consequently, the period which we may consider the most likely for the rise of vodka production was between the 1440s and the 1470s. The latest possible date when this might have occurred, we should note, was 1478. We are entitled to conclude that this was the year which saw the introduction of the vodka monopoly, since it was then that a state monopoly on foreign trade was introduced, and that legal steps were taken which introduced general financial control by the state over incomes from production and trade. Such products as salt and vodka, which were suitable for monopoly production and could readily be taxed, were undoubtedly the first to be monopolized.[12] This may have been before the final introduction of state control over trade (through the ousting of foreigners) or with the general establishment of control, as a final decisive measure. If these conclusions are correct, it follows that by the 1470s the production of vodka in the Muscovite state was already quite developed. Consequently the date of its rise must be sought some three or four decades earlier; that is, rather more than a human generation, since this is the minimum required for individuals, society and the state to acquire experience of the influence of vodka on various aspects of life.

In other words, we are talking about the years from the 1430s or 1440s to the 1470s. To refine and test this proposition, we shall examine a chart of the major developments in Moscow society during this period, since, as we established earlier, it is here that the emergence of vodka will be reflected. The chart includes natural events – fires, epidemics, famines and the like – since they too influenced the economic and social life of the population, lowering living standards.

1367: "Great Fire" of Moscow. The wooden Kremlin, the merchants' quarter and the Kitay-Gorod district are all consumed.

1367: outbreak of plague.

1375 : first execution in Moscow of a *tysiachskii* (elected military leader of the emergency volunteer corps).

1379 : annihilation by the Tatars at the river Pyana of Russian forces from Suzdal and Yaroslavl, as a result of mass drunkenness.

1389: baptism of the Permians after "peaceful" conquest.

1390: another large fire in Moscow. Several thousand houses reduced to ashes.

1395: yet another fire in Moscow. More than half the buildings constructed since the last fire are consumed.

1390–1424: constant epidemics of influenza and cholera throughout all the states of western Russia (Novgorod, Pskov, Smolensk and Tver).

1408–20: during these years epidemics affect the Moscow state only twice, and then are confined to its western and northern regions (Mozhaysk, Dmitrov, Yaroslavl, Kostroma and Rostov).

1410: Metropolitan Foty of Moscow issues an edict to the bishops introducing new rules of conduct. Priests are forbidden to engage in trade, to swear unseemly oaths, to marry anyone under the age of twelve, and to drink wine before dinner.

1421: snowless winter with heavy frosts. Meteorite showers.

1422: the only occasion on which an epidemic strikes the Moscow principality.

1424: the beginning of the new year is moved from 1 March to 1 September. The brewing of the ritual "March beer" in celebration of the pagan New Year (*iarila*) festival is abolished as far as possible. The Church begins an active campaign against beer–brewing as a pagan cult.

1425–6: transition from nicknames to surnames among the high-born and those employed in state service.

1426–31: plague is brought to Moscow from Livonia. *Oksymel* (mead vinegar) cannot be put to general use as a disinfectant because of its high price.

1427: over the fifty-five years since 1372 the population of Novgorod has shrunk by 80,000, counting only the people who have died in epidemics. In Moscow during the same period the population has doubled.

1430: the "Great Drought". Rivers dry up, and there are forest fires and "white haze".

1431: a census records the presence in Novgorod of 110,000 taxpayers. The republic's entire population, including people of all ages and the inhabitants of colonies, is around half a million people.

1437–9: first journey by a Moscow legation to Western Europe, to attend the eighth Ecumenical Council.

1442–8: further epidemics of plague throughout all the principalities.

1445: Moscow is looted and burnt by the Tatars, and the Grand Prince Vasily III is imprisoned.

1445: earthquake in Moscow. Dust storms.

1446: a rain of grain in Novgorod. The whole expanse between the Msta and Volkov rivers for fifteen *versty* (kilometres) is covered with a thick layer of grain. Evidently the grain was carried by storms from the fields of Moscow or Tver, since in Novgorod itself at that time there was a famine and the crops had perished.

1462: introduction in Moscow of "trade punishment" – the flogging in the marketplace of well-known citizens accused of
political crimes.

This list appears astonishing, or at least unexpected, in comparison with the list of economic developments over the same period.
It could rightly be called a catalogue of disasters; it might seem
to apply to a different, even a distant, period, so sharp is the contrast. One list details huge investments, giant construction projects, population growth and increasing crop yields; the other
records epidemics, natural disasters, famine and drastic population decline. How are we to reconcile these contradictory elements, and explain how they coexisted?

In the event, this is quite simple. The economic advances all
relate to the Moscow state, while the disasters occurred mainly
on the territory of the Novgorodian republic, if we exclude the
stroke of good fortune that provided Novgorod with a rain of
grain and saved many people from starvation.

Moreover, these natural calamities, when they extended on to
the territory of the Moscow state, did not produce there the same
devastatingly adverse effects as in Novgorod. First, there were
fewer of these disasters in Moscow, and second, they fell on less
troubled ground and were therefore overcome with comparative
ease. The most important events for Moscow during this period
were not the natural disasters which accelerated the economic
decline of Novgorod, but new social phenomena which bore
witness to changes in popular consciousness. Historians and contemporary observers alike record that during the fifteenth century, and especially between 1440 and 1460, an abrupt change
took place in morals and manners, which became coarsened to a
striking degree.

As testimony to this we find not only the introduction of new
types of state and social repression and their increased harshness
(severe punishments came in three varieties: the usual death
penalty, "trade punishment" and "religious" execution – that is,
burning alive), but also a changed relationship with the hierarchy of the religious estate, the public flogging of boyars, and the
introduction of surnames equally for boyars and for low-born but
literate state functionaries. Moreover, there was a drastic fall in

the prestige of the court *camarilla*, with the blinding of two princes, acts of treachery in public life and the routine poisoning of political opponents. The urban poor were provoked by trivial grievances to outbreaks of rioting during which the mob drowned and burned people without any form of trial. Of the same order was the breach of the formerly sacred inviolability of prisoners of war. During the first half of the fourteenth century there were frequent cases in which prisoners of war were subjected to vile humiliations.

There is a basis for linking these phenomena, if not completely then at least to a significant degree, with the dramatic spread of drunkenness; and above all with the change in the character of drunkenness, which now evoked not merriment but brutality. This too provides grounds for suggesting that changes had taken place in the character of alcoholic drinks.

The direction in which these changes were proceeding is quite clear. Production of mead, which in any case was destined for the tables of the nobility, was declining. The common masses had to be satisfied initially with a surrogate (the best known of these was a watery diluted mead or incompletely fermented mead vinegar infused with various narcotic herbs, including some that were more or less poisonous). Thanks to the outstanding ability of mead to mask the presence of various adulterants (even a seemingly insignificant admixture of honey or wax is enough to impart a mead-like aroma), this practice of faking alcoholic drinks became especially widespread during the fifteenth century, and in the sixteenth and seventeenth centuries it was aggravated by the habit of infusing both fake mead and vodka with tobacco, to give it "strength".

The spread of faked intoxicating liquors during the first half of the fifteenth century was stimulated by two factors in particular. The first of these was the growth of the population of Moscow and the Moscow principality and the concentration there of a large mass of seasonal workers. The second was the policy of the Church, which in 1424 began a systematic campaign to root out the domestic and collective brewing of beer. This coincided with the formal removal of the capital of All Russia from Vladimir to Moscow in 1426. From this time the coronations of Grand Princes, which previously had been performed in Vladimir, took place in Moscow. This formal shift in the centre of gravity of political life to the densely settled Muscovite capital and its environs strengthened still more the breach with the pagan traditions of beer-brewing, which were stronger in the Rostov-Suzdal region with its Finnic population than they were within the original boundaries of the Moscow state.

There is almost no doubt that the ban which Metropolitan Foty in 1410 placed on priests and monks drinking wine before dinner, even if this applied only to grape wine, was linked with the dwindling range of intoxicating liquors and the widespread consumption of bogus mead. As a consequence of this the monks and priests, who had access to supplies of official Church grape wine, began to squander it for their own delectation instead of employing it in the service of God.

It was this which moved Foty to instruct his bishops to reduce the consumption of wine. (Indirect evidence here is provided by the fact that Foty, a Greek who was named Metropolitan in 1409, was a native of Malvasia, the home of the best Greek wine imported into Russia. He could not have called vodka "wine". It follows that in 1410 such grain "wine" did not exist.) We can conclude that by 1410 the situation in the Moscow state with regard to the supply and consumption of alcoholic liquors had become difficult. It is clear that a desire must also have arisen during this period to find a drink that was cheaper than mead but just as intoxicating. The crusade by the clergy in 1424 against the brewing of beer provides still more evidence that grain spirit, that is, vodka, originated at some point very close to this time. One might even suggest that the Church would not have begun to campaign vigorously against the beer-brewing in the Moscow district if there had not been some alternative drink available. This is far from certain, but one should not exclude the possibility.

In the light of these circumstances the fact that in 1426 the Genoese, on their way to Lithuania, displayed *aqua vitae* to the Grand Prince and his court for the second time acquires weightier significance. It is entirely possible that on this occasion the display of *aqua vitae* aroused serious interest in producing it. Consequently, by the end of the 1420s or the beginning of the 1430s conditions for the production of grain spirit had ripened from all points of view, and above all from the angle of the existence of grain reserves. It was precisely during the 1430s that the results of the transition to the three-field system first became noticeable. At least from the beginning of the 1440s; that is, after a period of some nine to twelve years during which the three-field system gathered momentum, one can speak with near certainty of the existence of the reserves of grain needed for a successful start to vodka production.

Further evidence pointing to the 1440s as the period when alcohol distilling got under way in Russia is provided by the fact that at the end of the 1430s a Russian church legation visited Italy for the first time, in order to attend the eighth Ecumenical Council. It is known that the members of the legation had close contacts with the Catholic hierarchy of the Roman Curia. They also visited Italian monasteries and acquainted themselves with the organization of the Catholic orders and with the monastic economy, industries and way of life, since there was talk of union between Russian Orthodoxy and the Roman Church. It is not unlikely that while in the Italian monasteries the members of the Russian spiritual mission had the chance to acquaint themselves not only with *aqua vitae* as a product, but also to see the stills and to observe the process for themselves. This familiarity with equipment and techniques might have been of decisive significance for the initial organization of alcohol distilling in Russia since, as the saying goes, it is better to see something once for yourself than to hear about it a hundred times. No display of the drink *aqua vitae* could provide an idea of how alcoholic spirit was produced, but one glance at the equipment would have been enough to convince a visitor that this process was not complex. Among the Russian legation were people who were highly educated for their time: the Thessalian Greek Isidor, the Suzdal bishop Avraamy, and with them a hundred clerical and secular companions. They visited Rome, Venice, Florence and Ferrara.

On returning to Moscow Isidor was imprisoned in the Chudov monastery, where he spent a year before fleeing to Kiev and from there to Rome. It is astonishing first of all that he was not burnt by Vasily III for taking the side of the Roman Pope at the congress in Florence; second, that he was held in the Chudov monastery not as a criminal but under good conditions; third, that he was given a chance to escape unhindered, as well as being able to obtain transport and an escort; and fourth, that he was not pursued by Vasily III, but was given the chance to depart from Moscow state unharmed.

It is entirely possible that wishing to save his life, Isidor, an extremely astute Greek, succeeded in carrying out alcohol distilling on an experimental basis in the Chudov monastery; and that, not having any other raw material, he settled on the use of grain. If Isidor did indeed prepare alcoholic spirit, whose properties he knew well from his earlier journeys to Italy at the beginning of the 1430s, this might have made it easier for him to lull the guards to sleep and escape from the monastery.

Even if we consider this hypothesis totally unproven, as a suggestion it is completely legitimate since alcohol distilling in Russia could have originated and developed only in monastery laboratories, under the protection and patronage of the Church. This is why, even if alcohol distilling was not initiated in the Chudov monastery through the efforts of Isidor, we are justified in concluding that it was practised in some other large Moscow monastery at about this time, between 1440 and the 1460s or 1470s.

There are abundant facts to support this proposition. First, the monks in the privileged monasteries of Moscow were the most educated and technically accomplished people in the Moscow state. Second, they were familiar both from books and at first hand with the Byzantine practice of making *sikera* – the term was at this time used for a spirit made from raisins – and also could become acquainted either personally or through visiting Greeks (if they were not Greeks themselves) with the production of *aqua vitae* in Italy. Third, it was only in the conditions of a monastery that the necessary equipment could have been obtained and tested. Fourth, only the monastic hierarchy and the Church in the broader sense could have sanctioned such production.

Political and economic relations in the Moscow state between the Church and the secular power, the Grand Prince, were highly unusual and quite different from the relations beween the Church and the governments of Novgorod and Tver. The peculiarity of these relations in Moscow lay in the fact that the Moscow princes did not enter into conflict with the Church, and not only allowed it to take part in plundering the Russian people, but also refrained from arguing with it about the scale of the plunder.

While the Novgorod government intervened in petty fashion to ensure that the Church did not encroach too far on secular concerns, and while the prince of Tver forcibly prevented priests from engaging in trade, the Grand Princes of Moscow freed the Church, and especially its economic citadels, the monasteries, from all types of taxes, customs duties, tribute payments and other exactions of the feudal state. The Church was free from the direct "Tsar's duty", and also from paying duties on ploughs, post stations, horse-drawn vehicles or fodder, and other levies. The Moscow princes had learnt this policy from the Khans of the Golden Horde, who despite their Islamic faith were involved in an extremely close alliance with the Russian clergy. The results of this surprising and paradoxical union speak for themselves: throughout the whole of the fourteenth and fifteenth centuries in Russia there was not a single popular uprising against the Khans. In the churches the clergy prayed for the Khans as rulers of Russia!

Metropolitan Foty, a Greek who arrived in Russia in 1409 and who knew little of local conditions, dared to forbid the monasteries and priests from engaging in trade and from borrowing or lending money at interest. He met with such powerful opposition from the whole Church hierarchy that by 1411 he had shifted to a directly opposed policy – the taking back from lay citizens of monastery lands and property they had acquired during the previous years.

As powerful landowners the monasteries also gained most of the benefit from the transition to the three-field system. This innovation had been borrowed by the Church from Byzantium and Greece with their developed agriculture, and was therefore unknown to Russia's northern European neighbours, which did not have links with the orthodox East. For economic, historical and technical reasons, therefore, the monasteries were the most likely places for alcohol distilling to arise in Russian conditions. Here was to be found everything that was lacking both among the boyars, and to an even greater extent among other social classes: raw materials, personnel, knowledge, equipment, and a secure power base. Moreover, the Church had clear economic and political aims which were furthered by alcohol distilling.

In economic terms the Church was not only well disposed to opening up a rich vein of finance, but also had experience of the monopoly exploitation of such a source of income. In Kievan Russia the monopoly on salt – the most ancient of all monopolies – had been held not by the princes or the government but by the Church, and specifically by the Kievo-Pechersky monastery. It is quite possible that the monasteries hoped to seize the monopoly on vodka as well.

The factors which played a role here were not limited to economic considerations, important as these were. Since the middle of the fifteenth century the Church's aims had included the conquest of eastern territories inhabited by pagans. Therefore the Church, or at least its individual representatives, might well have thought to speed the conversion of these pagans by employing the "miraculous and irresistible properties" of vodka. If Stefan Permsky (1345–96) had not yet been able to make use of this expedient before he performed the baptism of the Komi-Permians between 1379 and 1383, his successors, and especially Pitirim, used alcohol to help achieve the baptism of the warlike Mansi. Since Pitirim's missionary activity was carried out between 1447 and 1455 (when he was killed by the Mansi leader Asyk) there is reason to suppose that the monks had initiated production of grain spirit at least by the end of the 1440s, and perhaps somewhat earlier.

Thus, a series of diverse and indirect pieces of evidence leads us step by step to the conclusion that the distillation of alcohol arose in the Moscow state, and most likely in one of the monasteries of Moscow itself, perhaps in the Chudov monastery, in the period from the 1440s to the 1470s. The year 1478 can be considered the terminal date, when the production of grain spirit had already been proceeding for some time and when experience of this industry led the state to introduce a monopoly on the production and sale of vodka.

However, if a conclusion reached on the basis of this indirect historical evidence is to be fully convincing, one needs to answer the inevitable question: if grain spirit was indeed prepared in Russian monasteries, and in those of Moscow in particular, during the second half of the fifteenth century, how could it be that no report of this has survived, either in the Russian chronicles which were compiled and copied by monastery scribes during the fifteenth century (and especially during its second half), or in the monasteries' financial documents of the fifteenth and early sixteenth centuries?

Does not the absence of documentary material prove that vodka production arose either in another place or at another time, and that as a result, the conclusion reached on the basis of our research cannot be considered irrefutable?

Why Do the Chronicles and Account Books Tell Us Nothing?

Even if the time when grain spirit production began was considered unworthy of inclusion in the chronicles, then by its very properties vodka or "burning wine" must have struck the imagination of the people of the period, and in one way or another they would have expressed their views on this novelty which was having such an effect on morals and social conduct. But the chronicles are silent. Moreover, even passing hints at the existence of grain spirit are totally lacking from these records throughout the periods which they cover. This is despite the fact that most of the scribes who copied the chronicles were alive during the middle or the second half of the fifteenth century.

How do we explain this mystery? First of all, we have to consider what the Russian chronicles were. How and when did they come into being, and who were the Russian chroniclers?

The oldest of the chronicles begins its record of events from the ninth century, and the accounts are carried through approximately to the middle of the fifteenth century. The events of the second half of the fifteenth century are, however, left out. This is the case with all the Russian chronicles, and with all their versions and variants, including the lists compiled during the second half of the fifteenth century, in 1456, 1472, 1479 and 1494. These latter summaries added not a word about what had happened during the second half of the fifteenth century, the period when the chroniclers themselves were alive and to which they were eyewitnesses.

In their basic content the chronicles include facts only up to 1423. A few variants record what happened in Russia in general and Moscow state in particular up to 1431 or, in the case of the fullest version, to 1448. Only such separate regional records as the Dvina, Vyatka and Perm chronicles touch on events of the second half of the fifteenth century, and then in an extremely fragmentary and unsystematic manner.

Consequently, if we were to consider that the chroniclers could not have left out such developments as the beginning of alcohol distillation or the introduction of the vodka monopoly, this would mean that such events did not occur at the latest until 1448, or perhaps until the period from 1423 to 1431.

Such a conclusion, however, cannot be embraced too whole-heartedly, even though it is logical and might define more pre-cisely the date when alcohol distilling made its appearance. The Russian chronicles, it turns out, have two peculiarities which suggest that they might well simply have ignored the rise of vodka production, even if it occurred before 1448.

First, the Russian chronicles report the facts of economic his-tory in a notably brief and incomplete fashion, even if these facts are extremely important. This was noted by A. Shletser, the first historian to perform research into the chronicles. The compilers of these documents might record punctiliously the details of a fight in a street or square of Moscow or Novgorod, since such incidents struck them as important; but would say nothing of the conclusion of a trade agreement, or of the goods which were to be found on sale in Moscow. We are obliged to seek out such information from foreign sources. It is not that alcohol is not mentioned: the chroniclers could give a meticulous account of a drunken brawl, or on drunkenness in general, but still fail to note the advent of vodka distilling.

Second, and in contrast to our views today, medieval chroniclers considered it necessary to recount events from the distant past, which they had not themselves witnessed, in greater detail than recent or contemporary events, which they did not consider should be recorded at all. As a result it is clear that the people who com-piled the chronicles in the second half of the fifteenth century would not have reported a single one of the events of this period. If alcohol distilling made its appearance after 1448, it follows, we would never find a single reference to this in the chronicles. These documents were meant as archives, and it seemed perfectly reason-able to the compilers not to include contemporary material.[13]

The absence from the chronicles of direct references to the appearance in Russia of grain spirit production is therefore to be explained by the facts that, first, the chronicles did not deal with the events of the second half of the fifteenth century, and secondly, developments in the field of economic history could be omitted from the chronicles even if they took place earlier than the middle of the fifteenth century.

However, the rise of distilling should have been reflected in the financial documents of the monasteries and of the palace administration, and in documents of the local offices, the "interior ministry" of the period. Unfortunately, documents of this type have not been preserved to any significant extent. The economic documents in our archives date mainly from the seventeenth century, and only a very small number touch on events during the late sixteenth century. A survey of monastery records, which might reveal not only the date of appearance of grain spirit but also the first steps in the development of the technology of Russian vodka, shows that, although many monasteries arose long before the fifteenth century, they did not subsequently retain documents from their early history; that is, from the period in which we are interested. The following table shows this clearly.

Survival of monastery records

Date of founding of monastery	Name of monastery	Year from which earliest surviving records date
12th century	St Anthony of Novgorod	1653
12th century	St Nikita of Pereyaslavl	1617
12th century	St Nicholas Gostinnopolsky	1623
13th century	St Nicholas Kosinsky (Staraya Russa)	1660
13th century	Staraya Ladoga	1662
1335	Resurrection of Novgorod	1620
1330	St Hypation of Kostroma	1596
1330	Monastery of the Caves (Nizhny Novgorod)	1615
1337	Trinity of St Sergius (Zagorsk)	1466
1365	Protection of the Virgin (Suzdal)	1587
1371	The Saviour of Priluka (Vologda)	1510

Survival of monastery records – continued

Date of founding of monastery	Name of monastery	Year from which earliest surviving records date
1371	St Aleksey of Uglich	1639
1374	Vysotsky of Serpukhov	1677
1393	Ascension of Moscow	1691
1397	St Cyril of Beloozero	1522
1398	St Therapont of Beloozero	1438
1420	Stavro-Voznesensky (Pskov)	1628
1426	Bogoslovsky Vazhsky	1588
1479	St Joseph of Volokolamsk	1532
1490	St Anthony of Siya	1490

From this table it is evident that the monasteries of Moscow retained no archival materials of use to us, since their repositories were almost completely destroyed during the period of Polish-Swedish invasion and peasant wars between 1604 and 1612. The archives of monasteries remote from Moscow were preserved better, but only the St Therapont and Trinity of St Sergius monasteries retained documents from the second half of the fifteenth century. Unfortunately, these relate only to the monasteries' political history; they consist of deeds granting ownership, or relating to grants bestowed in return for services in the field of defence, or recording loans, or similar matters.

It is important to note, however, that the monastery documents of a later period do indeed confirm that grain spirit was being produced on the premises. Testifying to this, for example, are accounts for expenditures on raw materials in monastery breweries (1587), documents relating to the dismissal of monastery liquor-sellers, and the final edicts of Peter I concerning the handing over to the monasteries of all copper stills and vats. It is quite clear that the monasteries could have produced vodka under the conditions of the monopoly only if they had begun this production before the monopoly was instituted. The hypothesis that alcohol distilling in Russia arose within monasteries is thus confirmed, though in indirect fashion.

It needs to be explained, however, why it was that Russian

distilling and its end result, vodka, were not reflected in the later monastery literature which has survived, in the records of the lives of saints and of ordinary monks.

The answer to this question is quite simple. Who were the chroniclers, the inventors, the writers, the moralists and ideologues of the Middle Ages? In the overwhelming majority of cases they were monks. During the fourteenth and fifteenth centuries this fraternity was not so idle and useless as it notoriously was in the nineteenth century. Monks played diverse roles in the medieval state. They included scientists and technicians, inventors and alchemists, with their laboratories kept secret from the monasteries' other inhabitants. There were philosophers and chroniclers, often acting as advisers to the Grand Prince or compiling their chronicles in line with the will of the monarch; or conversely, slanting their records to portray the monarch in a bad light. It stands to reason that if the first experiments in distillation were carried on within the walls of the monasteries, only monks could have reported on them. The monks could have done this, but for many reasons might not have wished to do so.

The discovery of the drink might have been recorded in secret monastery documents which have not come down to our time, but it is highly probable that this information did not appear in any later Church literature for purely ideological reasons. Within forty or fifty years of the appearance of vodka the baneful social and moral consequences of this discovery had become obvious. Now that vodka had become the principal means of replenishing the state treasury, and all other people and institutions, including the monasteries, had been forbidden from selling it, the Church emerged as one of the drink's most bitter opponents. To be more precise, the object of clerical wrath was not vodka itself but its consequence – drunkenness. In these circumstances the monks and clerics in general were forced to conceal the fact of their involvement in creating the "devilish poison". There even appeared a folk tale relating how the Devil had taught a peasant how to make vodka. This particular tale arose very late, at the beginning of the nineteenth century, but similar explanations were put about by the Church much earlier.

All of this is additional evidence to explain the mystery with which the date of the origin of vodka had come to be surrounded, the complete disppearance of documents relating to the rise of distilling, and the still more striking lack of any mention of the beginning of vodka production in the historical materials relating to the monasteries during the time when many monastic archives were still intact and accessible.

In all probability, the internal monastery documents relating to the discovery, technology and production of vodka were assiduously destroyed in the middle years of the seventeenth century during the period of struggle against the followers of Nikon and the schismatics, and especially after the arrest and exile of Nikon himself. This was done in order to deny weapons to ideological opponents, and in general, to make a decisive break with all the "sins" which the Church had committed prior to its reformation. A furious struggle was waged over the "correcting of the old entries in the books of the Church".

It would be naive to think that the only passages which were deleted or "corrected" were those which mentioned the use of two fingers in performing the sign of the cross, or similar trifles and formalities. Friedrich Engels observed wittily at one point that if the theorem of the isosceles triangle had somehow touched on human economic interests, wars would have been fought over how to prove it.

Around the middle of the seventeenth century, following the vodka riot that exploded into the Stenka Razin uprising, the Church was forbidden to practise alcohol distilling. This is reflected in the surviving documents from the second half of the seventeenth century, when all matters relating to alcoholic spirits were transferred to the control of the local administrative offices.

Around the end of the seventeenth century some of the monasteries again began distilling vodka for private sale, but these ventures were again cut short by the state in 1719. No longer permitted to draw profits from vodka, the Church, and especially the monasteries, became its "natural" adversaries in the fields both of ideology and economics. This explains the fact that nowhere in medieval or later sources relating to the history of the Russian Church, nor in the researches of pre-Revolutionary historians, is there any mention of monasteries as the birthplace of alcohol distilling in the Russian state.

Moreover, there is no mention of alcohol distilling in the list of manufactures and trades carried on in the monasteries during the period from the fifteenth to the seventeenth century, as set down in monastic records. Recorded activities are fishing, the extraction of salt from brine, the production of wooden bowls and dishes, cooperage, haymaking, malting, brickmaking, pottery, stonecutting, the extraction of saltpetre, dairying, the preparation of bark (from oaks, alders and birches), leather dressing, the production of scalded cream, tallow rendering, gunpowder production, the casting of cannon and cannonballs, flour milling, timber cutting and carting.

The absence of alcohol distilling is understandable: the list enumerates only the trades and manufactures carried on in the northern, not in the Moscow monasteries. Among these pursuits is malting; that is, the production of malt, and hence of beer, *kvas* and ale. As was shown in the previous chapter, in technical terms these processes were quite unrelated to alcohol distilling, and consequently, where the brewing of beer was carried on in the fifteenth century the production of grain spirit was almost automatically absent. In cases where the monastery economy included an enterprise for preparing alcoholic drinks, we can assume that the malthouses in the northern monasteries with their Novgorod traditions were replaced by distilleries in the monasteries of Moscow and regions to the west.

Evidence that distilling was still carried on secretly in these monasteries is provided by isolated documents of the Monastery of St Nicholas in Koryazhemsk and of the Monastery of the Protection of the Virgin in Suzdal, although these documents relate to a later period. From these sources it is clear that the monasteries were not only secretly producing vodka but also secretly selling it, and sometimes also buying it from outside. The monastery hierarchs, discovering this trade when investigating fraudulent accounting in the taverns, sacked and replaced the liquor-sellers in the monastery breweries.

It is interesting to note that the early seventeenth century saw the flight in 1615 from the Monastery of the Caves in Nizhny Novgorod of the *bortniki*; that is, of the people whose work was to gather honey from wild bees and to use it to produce matured mead. It is obvious that by this time either their conditions of work had become extremely burdensome, or else their now rare skills were more esteemed and better paid outside the monastery. In any case it is indisputable that production of grain spirit had begun in the monasteries and that alcohol distilling had been transformed into one of the trades of the monastery economy; at least before the beginning of the sixteenth century, when the vodka monopoly entered into full force. It is also apparent that despite being well established this fact not only went unrecorded in the historical literature, but was deliberately concealed by the clerics.

Finally, the example of the monasteries also allows us to follow the trend in the preparation of alcoholic beverages which had its beginning in the middle of the fifteenth century and which was general for all Russia: a sharp territorial demarcation of beer-brewing and vodka-distilling regions, combined with a general contraction of brewing and growth of distilling, and the gradual decline and disappearance by the seventeenth century of the production and maturing of mead.

Fixing the Date

What conclusions can we draw from all the materials we have examined? Which period during the fifteenth century should be considered most likely to have seen the rise of distilling?

Earlier we noted that of the three periods into which we divided the fifteenth century – that is, from 1399 to 1453, from 1453 to 1472, and from 1472 to 1506 – the most likely to have witnessed the emergence of distilling, judging by our evaluation of the history of the Muscovite state, was the last of these. If we consider the indications that the rapid economic growth and development of financial activity during these years had been stimulated by alcohol distilling, then we can move the threshold of this period at least a decade back, and conclude that vodka made its appearance somewhere between 1460 and 1500.

Our analysis of the economic material has shown that the transition to the three-field system which occurred during the 1420s and 1430s was an important factor in the creation of grain surpluses which were used as the raw materials for grain spirit production, and that without these surpluses the transition to large-scale distilling as a state-run sector of the economy would have been impossible. Distilling, if it had been discovered, would necessarily have remained at the experimental stage for a prolonged period thereafter.

On the basis of the economic material, therefore, we can move the date for the beginning of vodka production farther along to the period from the mid-1420s to the early 1430s.

At the same time, the evidence of economic history shows clearly that by 1478 the Russian market was becoming consolidated to such an extent that the government was establishing *de facto* control over it, rejecting the services of foreign merchants as intermediaries in both the domestic and foreign markets. One is entitled to conclude that this step was taken either in anticipation of a monopoly on vodka, or immediately following the introduction of such a monopoly. This means that by 1478 or 1480 the production of vodka was an established fact, and that it had reached a more or less stable level. The economic evidence thus allows us to conclude that vodka made its appearance in Russia at some time between 1430 and 1480.

An analysis of social factors confirms these dates in general, but does not provide a completely clear picture.

The struggle by the Church against the brewing of beer and against the pagan cult of drunkenness, celebrated over strictly defined but always prolonged periods lasting from three days to two weeks, provides indirect testimony to the beginning of grain spirit production; the only drink that could be counterposed to beer in 1425 was vodka. The ease with which this new drink could be portioned and packaged and its uniform quality gave vodka a cosmopolitan character, making it an everyday commodity without sacred associations. Vodka had no links to any particular events, and had emerged through an objective historical process as an agent of non-traditional drunkenness. At the time, this characteristic of vodka appeared as an important distinguishing property both in the view of the Church, since vodka stood in contrast to the pagan drinks, beer and mead; and in that of the state administration, which was anxious to bring to an end universal intoxication at particular times of the year. Under the conditions of the money economy and with a relatively large population, this custom was destroying both administrative order and the rhythm of economic life for long periods.

It had also been observed that ritual drunkenness and the

drinking of beer aggravated epidemics and aided their spread, while the use of vodka tended to limit them.[14] This was especially true of epidemics of respiratory diseases, which at that time were not distinguished from plague and other infections. The disease which raged in Novgorod almost every year from 1408 to 1422 was not plague, but a strain of influenza.

On the basis of the fact that the pandemic of respiratory disease at that time did not affect Moscow, one could argue that from the middle of the 1420s, and perhaps from the beginning of the fifteenth century, vodka was already known there. The peak of economic prosperity and the largest harvests came at the beginning of the 1440s, and it was precisely these food surpluses which formed the indispensable base for the development of distilling. Meanwhile, the abrupt decline of morals was most clearly expressed from the middle of the 1440s.

Hence the social and economic material taken as a whole allows us to settle on the period from 1440 to 1480 as the most likely to have witnessed the birth of vodka distilling. Of course, it is not impossible that this development might have occurred in the years between 1425 and 1440, but we have no substantial evidence to this effect. Weighed against this possibility is the fact that vodka is not mentioned in any of the chronicles that finished with the middle of the fifteenth century, and in which, consequently, the events of the period between 1425 and 1440 were recorded, in some cases either by eyewitnesses or by people from the next generation who had known them. This would have guaranteed that the beginning of distilling or the emergence of vodka, if such developments had occurred during living memory, would have been included in the chronicles.

The facts that no such events are recorded, and that the latest chronicle finishes in 1448, lead us to conclude that in Russia distilling and the production of vodka arose between 1448 and 1478. This is the period to which the evidence points most clearly, and which is least open to objection. While a date earlier than the middle of the fifteenth century, perhaps in the period between 1425 and 1440 and conceivably somewhere around the divide between the fourteenth and fifteenth century, cannot be excluded, there is little evidence for it. A date between 1448 and 1478 is supported by the whole sum of historical, economic and social research. We may take it no longer as a hypothesis but as a logical conclusion.

Now that the period has been narrowed to thirty years, it would be possibile to search for a more precise date. However, the tasks of the present work do not include establishing this date with minute exactitude. The years between 1448 and 1478 represent a sufficiently narrow span for dating the appearance and development of such a type of industrial production. Furthermore, the fact that by 1478 an official monopoly had been imposed on the production of grain spirit indicates that this production was not only well developed, but that the commodity itself had acquired a standard character and conformed to an accepted level of quality. Vodka, at this point still known as "burning wine", was legally distinguished by the state from all types of imitations and from the home brews which had received the official description of *korchma*.

It is notable that distillation of grain alcohol made its appearance in the Moscow state earlier than in nearby countries such as Denmark, the German states, Sweden, Poland and Moldavia, not to speak of the other states making up Russia at that time. In Sweden, for example, to which one group of historians considers that vodka was first imported from Russia in 1505, while another group insists that production began inside the country earlier than this, an effort has been made to establish the date of appearance of vodka through calculating the consumption of salt by the population. Since a salt tax was in force in Sweden in the fifteenth and sixteenth centuries and lists of payments of this tax have been preserved for every village and every family, it is easy to work out how the consumption of salt altered over time. The years when the consumption of salt rose most sharply have been suggested by Swedish historians as those when alcohol distilling made its appearance. The reasoning is that people who eat salted food drink more! In historical terms such a method, however precisely based on mathematical calculations, is primitive and implausible.

To compare the development of alcohol distilling in Russia with the rise of this industry in neighbouring countries, a chronological table is useful. From this table is clear, first, that the development of alcohol distilling and the whole cycle of state regulatory activity, from the imposition of a ban to the granting of

The development of distilling in Russia, Sweden and Germany

	Russia	Sweden	Germany
First mention of a sample of alcoholic spirit, or first written account of how to obtain it	1386 (imported sample)	1446 (experimental production of a sample)	1512 (translation from Latin to German of a book on distilling)
Beginning of sporadic domestic production	1399–1410	1490	1525–1528
Ban on the production of vodka	1425	1550	1545
Production in monasteries	1440?	?	?
Unauthorized distribution, demand for removal of ban, or official removal of ban	1440?	1560	1612
Introduction of monopoly	1478	1638	1648

permission to the placing of a monopoly on vodka production, occurred in Russia approximately a century earlier than in neighbouring countries. Secondly, in all these countries the development of alcohol distilling was a complex and prolonged process which extended over some fifty to a hundred years and was marked by interruptions. Russia was no exception, but it was here that this process went ahead most rapidly, with the introduction of a monopoly on grain spirit production preceding the same move in Germany by nearly two hundred years.

We have here a superb illustration of how strongly centralized Russia had become, and of the disunity and decentralization of medieval Germany, since the introduction of a monopoly is a classic indicator of a high degree of centralization of the state.

We can now turn to the history of vodka production from the middle of the fifteenth to the middle of the nineteenth century, that is, to the point at which it was placed on a modern industrial footing and a new stage in its technology began. Our task here will be to trace in detail how grain spirit became transformed into vodka; that is, into a product endowed with special qualities that were stipulated at a national level. We shall describe the stages through which the development of vodka passed, the types of vodka which were produced, and the technical and other improvements which established the quality of the drink and influenced its final character.

This information, while part of the realm of history, can also have considerable importance for improving and ultimately perfecting the quality of today's vodka. A knowledge of the history of vodka may well make it possible to reintroduce into the modern formula refinements which have been forgotten or have been considered unsuitable for use within the framework of factory production, but which are capable of improving vodka and also of renewing and restoring it. This knowledge may also make possible the recreation of a number of forgotten historical varieties of vodka.

With this in view we need to examine the terminology that was used with relation to vodka from the mid-fifteenth to the mid-nineteenth centuries, and to reveal the meaning of these terms.

3

The Terminology of Grain Spirit from the Fifteenth to the Nineteenth Century

Distinguishing Grain from Grape

From the very beginning of its production until the 1860s, and even for some decades after this, grain spirit was known by the term *vino*, also used for grape wine. Grain *vino*, however, was referred to by qualifying epithets which were different from those used for grape wine, and also by a number of additional terms in which the word *vino* was not included. Between the fifteenth and nineteenth centuries several terms for grain spirit were current. In essence, these were identical in meaning to *vodka*, although this term arose much later. If we are to identify vodka during the various historical periods of its existence, it is thus indispensable to know all the terms or nicknames by which it was known during the periods in question. Very often, several different names for vodka were current at the same time.

From the context of a particular documentary source, whether a literary text or a commercial or financial record, it is usually quite easy to work out which kind of "wine" – grape wine or grain spirit – is being discussed. There are various pointers to this: the way in which the drinks are consumed, price, volume measurements, containers and so forth. But when the reference to the drinks is outside such a context, or when the context provides few additional descriptive signs, researchers now find it extremely difficult to find their way among liquor-trade or distilling terms which have long since fallen into disuse. It is still harder to say with any precision what each term means and to which period it relates. For this reason it is extremely important to examine the terminology of alcoholic liquors on a chronological basis, juxtaposing terms signifying "vodka" and "grain wine" in each period with those used for grape wine. In this way confusion between these two categories can be avoided. Together with commercial terms and those relating to the everyday consumption of alcoholic liquors, we shall list separately a different category of terms concerned with the degree to which these drinks were processed.

Once we are familiar with both terminological systems, we shall have a reasonably complete idea both of the way in which the nomenclature of vodka evolved over the centuries, and of the changes in the composition of the drink, its quality, and the techniques of its production.

The Terminology of Grape Wine from the Fifteenth to the Eighteenth Century

In Chapter 1 it was noted that the term *vino* in Russian as well as in Old Slavonic was a direct borrowing from Greek, and was introduced into everyday language through the translation of the New Testament at the end of the ninth century. With the appearance in the fifteenth century of grain spirit the terminology of grape wine became more complex and detailed. To the terms *ots'tno vino* (sour, dry) and *osmir'neno vino* (sweet, dessert) were added others which distinguished between grape wines on the basis of their colour: *krasnoe vino* (red wine, 1423) and *beloe vino* (white wine, 1534). Alongside the old term "church wine", from the sixteenth century we find *vino sluzhebnoe* (service wine, 1592). Also during the sixteenth century people began to differentiate between grape wines on the basis of their maturity. Thus *vino vetkhoe* (old wine) meant wine that was matured and of superior quality. Ordinary wines began (1526–42) to be called *vino nerastvorennoe* ("undiluted" wines) to distinguish them from wines which were consigned directly to the table, and which therefore were diluted with water following the Greek custom.

From the middle of the sixteenth century, and especially from the seventeenth century when winemaking appeared for the first time in Russia,[1] grape wines were generally known according to their country of origin. The name *friazhskie* (Frankish) was at first applied indiscriminately to French and Italian wines. *Ugorskoe* (Hungarian) wine referred, as a rule, to various types of Tokay. The adjective *renskoe* (Rhenish) was applied to German white wines (1680). Wines from Greece and Asia Minor, meanwhile, went under exclusively local names: *kosskoe, bastro* (1550–70), *malvaziia* (1509–20), and *kiprskoe*.

French wines, for example, Romanée and muscatel (1550), were also described on the basis of their place names or the grapes they were made from, from the end of the fifteenth century. In the seventeenth century Spanish and Sicilian names – Jerez, Málaga and Marsala – made their way into the language.

From 1731 wines as understood in the West began perforce to be called "grape" wines, in addition to the other terms that were applied to them. This was particularly the case where they were mentioned officially in trade; there they were described as "red grape wine", "white grape wine", "malaga grape wine" and so forth. It was from this time, in connection with the development of a more detailed and precise terminology for grape wines, that the term *vodka* also began to be used to designate the spirits produced from these wines. Thus from 1731 armagnac and cognac were called "French vodka", while Russian fruit vodka was termed *kizliarskaia vodka*.

Trade and Everyday Terms for Grain Spirit from the Fifteenth to the Nineteenth Century

Khlebnoe vino (grain wine): this term is first recorded in a document of 1653. From the second half of the seventeenth century it was the usual literary expression for vodka or grain spirit. The term was not, however, used in official documents; in all trade, administrative and legal documents from 1659 to 1765 only the word *vino* appears, without any epithet. From 1765 an obligatory qualifier, referring to the degree of purification, was added in official references; together with the introduction in 1731 of a precise terminology for grape wines, this made it possible to distinguish clearly between references to vodka and to grape wine. Until this time various terms had been used in unofficial contexts, in daily life, in literature and in trade – especially retail trade. These terms singled out grain spirit, serving either as an equivalent or as a substitute for the qualifier "grain", as a euphemism or metonym; or, from the second half of the eighteenth century, as a direct indication of the degree of refinement of the product.

Meanwhile, until 1649 and to some degree until 1765 archaic terms remained in customary use, as they had been assimilated into the speech of the common people or employed in old documents. The oldest of these terms were *varenoe vino* (distilled wine), *korchma* (hooch; literally pot wine) and *goriashchee vino* (burning wine).

Varenoe vino (distilled wine): this was one of the first, and perhaps the very first, of the terms linked with the production of vodka. Its antiquity is established by a document of 1399 which contains the phrase "drinking distilled wine", but the most important thing to note is that semantically the term "distilled wine" was created on the basis of analogy with the terms "brewed mead" and "brewed beer". A term which was close to this one is *perevara*, which was in use from the middle of the fourteenth to the beginning of the sixteenth century. This denoted a strong drink prepared by means of heating low-quality brewed mead together with beer. It is clear that *perevara* was, in an indirect way, a precursor of "distilled wine" or vodka. It was *perevara*, as a substitute for good-quality mead, that was the principal agent responsible for the events on the river Pyana in 1377 and the capture of Moscow by Tokhtamysh in 1382. This mixture of two different drinks gave rise to the idea of strengthening the intoxicating effects of brewed mead by heating it together with a ready-made alcoholic product containing malt and grain. The high concentration of alcoholic congeners that resulted from the mixing of these ingredients, and the strongly intoxicating effects that resulted, were attributed exclusively to the grain component. This encouraged people to use grain on its own as a raw material for distilled drinks. This explanation of the rise of vodka successfully addresses all the apparently contradictory facts bearing on this question, to the point where the contradiction disappears.

A new, cheap and powerful intoxicant, *perevara* is mentioned as a trade commodity in the agreement of 1470 between Moscow and Lithuania. The term occurs for the last time in an official document of 1495. Relatively soon afterwards, and certainly no later than 1530, production of *perevara* ceased completely, and the place of this drink was taken by "distilled wine"; the latter phrase was undoubtedly one of the first, and perhaps the very first official term for vodka. It is possible, however, that for a time "distilled wine" was known as *perevara*, since the meaning of the term "distilled wine" was not especially clear. At the same time, the succession of terms beginning with the old product, brewed mead, and proceeding through a transitional, intermediate type of product, *perevara*, to a new product and process, "distilled wine" originating from true distillation, is beyond question. It testifies convincingly to the fact that "distilled wine" was the first term for vodka, at a time when the process of distilling was not yet recognized as a qualitatively new technique.

Korchma (pot wine): this was one of the oldest official or semi-official terms for vodka, existing almost contemporaneously with the term "distilled wine", but outliving it by almost four centuries. *Korchma* originally signified home-made vodka, distilled by people in their own households for personal use and for that of guests and neighbours. Later, after the state monopoly on vodka had been introduced in the fifteenth century, the term *korchma* came to denote illegally produced vodka, such as would today be called *samogon*.

The term *korchma* in the latter meaning is older than *vodka*, and before the discovery of vodka was applied in the same sense to mead, beer and even, in the eighteenth century, to *kvas*, when any of these drinks was sold illegally. As early as 1397 one of the city ordinances of Pskov decreed that townsfolk could not "keep *korchma*, or sell mead by the barrel". Thus the sale and consequently also the production of *korchma* was restricted, though not in a consistent fashion.[2]

In 1474 we find a still more definite instruction: "*Korchma* must not be brought into Pskov, or bought and sold."[3] The drink referred to here was undoubtedly vodka, since we are told that in 1471 "all types of goods could be had in abundance in Novgorod, and grain was cheap".[4]

In the legal code of Ivan IV ("the Terrible") and in the Code of 1649, *korchemstvo* is defined as the illicit, clandestine production and sale of vodka. This term remained current, with the same meaning, throughout the eighteenth century.[5] A special statute relating to *korchma* was even adopted in 1751, and periodically amended until 1785. This related to tavern-keepers who sold *korchma*, and provided for keeping watch for illicit liquor (*korchemnoe vino*) at town gates, for the labelling of barrels, and for the eradication of illicit distilling in Moscow and St Petersburg.[6]

As can be seen from these examples, during the eighteenth century the term *korchemstvo* was used to refer to any illicit sale of any liquor, and not only vodka. But during the fifteenth and sixteenth centuries, and even during the first half of the seventeenth century, the terms *korchma* and *korchemnoe vino* were linked exclusively with vodka. Thus, in the treaty between Pskov and the Hanseatic League at the end of the fifteenth century it was set out that German merchants could not import "beer and *korchma*" into Pskov. This not only provides an indirect indication that *korchma* corresponded in meaning to vodka, but also furnishes additional confirmation that a monopoly of the trade in alcoholic spirits was in force in the mid-1470s.

Kurenoe vino (a more precise term for distilled wine): this expression is first encountered in the Charter of the Monastery of St Joseph of Volok in 1515, and in all probability follows chronologically after the two preceding terms. It is also recorded in the seventeenth century, in 1632. But it is apparent that in the second half of the seventeenth century the term was no longer used. This is quite understandable, since it referred only to the technical peculiarities of the new process of grain alcohol distilling, and not to the characteristics of vodka so far as consumers were concerned. Vodka was most likely known under this term mainly to a narrow circle of its producers, to those concerned with storing and guarding it, and to the administrative, fiscal and police authorities. Hence this exclusively technical term did not endure for long.

Goriachee vino (burning wine, 1653); also *goriuchee, goriushchee, goreloe* (1672); *zhzhenoe vino* (burnt wine, 1672): this term in its several variants was used extremely widely during the seventeenth century, also persisting throughout the eighteenth century and to some extent into the nineteenth and even up until the twentieth century. It appears in official state documents in the forms *goriachitelnye* and *goriachie napitki* (burning drinks).[7] Despite

its wide distribution and prolonged use the term was very unstable and became profoundly distorted. The most correct form is
goriashchee vino, that is, wine which has the property of burning,
and not *goriachee*, having a high temperature. Still less satisfactory
is *goreloe*, burnt.

This term is essentially a pan-European one, since in the form
Branntwein it was adopted in all German-speaking countries as the
generic name for spirits, including vodka. However the German
term does literally mean "burnt", referring to the heat of the distilling process. In the Ukrainian language the term, in the form
gorilka, became the main official designation for vodka, and in
Poland it became one of the two principal names, *gorzalka* and
wódka. *Gorzalka* is used in spoken Polish more often than *wódka*,
which predominates on labels and in advertisements.

In the Russian language, however, *goriashchee vino* (or *goriachee*,
goreloe) did not become either the basic or the predominant term,
despite its wide distribution in the seventeenth and eighteenth
centuries, and it ultimately failed to become established in popular
consciousness or in the language. The reason lies above all in the
inability of the common people to understand the essence of the
term, and in the diverse ways in which it was interpreted by different groups within the population. As a result it could not
become the general national term, in contrast to the Ukraine,
where the word *gorilka*, (which means "flaming", from *goriti*, "to
blaze up"), could only be understood in one way and would not
mislead. But in the Russian language, with *goriachee* and *goreloe*,
the situation was quite different. This was the main reason for the
term not becoming established in Russia. In addition, it was an
imported, translated term, not Russian in character. The Russians
were less inclined to identify vodka (or any other product) on the
basis of its physical or technical characteristics than on the basis
of the raw material from which it was made, the peculiarities of
the technology, or the place of origin.

This last consideration was especially characteristic of the
Russian psychology, and remains an important distinguishing feature right up to the present day. We prefer to say "Dutch cheese",
"Yugoslav plums" or "Hungarian chickens" instead of giving
them generic names such as Edam, *reineclaude* or Leghorn, as
other nations would. Our usage makes it immediately clear where

each item is from and what it represents. In exactly the same way
a person of the period from the fifteenth to the seventeenth centuries habitually said "Russian wine", "Cherkassk wine" or
"French wine", and understood perfectly that the first two terms
referred to vodka, while the third referred to grape wine, although
there was no grammatical difference.

Russkoe vino (Russian wine): this term arose in the second half
of the seventeenth century, and a source records it in 1667.
Instances of its use are encountered relatively rarely, but it figured
in official foreign trade documents. The term reflects vodka producers' clear consciousness of the drink's national characteristics,
of its inimitable taste and quality, and of its special properties
compared with other types of the same product such as "Lifland
wine" and "Cherkassk wine".

Lifliandskoe vino (Lifland wine, 1667): this term was current in
the seventeenth century in the lands adjoining the Baltic region,
in Pskov and Novgorod. It was applied to grain spirit imported
from Estonia and from Riga.

Cherkasskoe vino (Cherkassk wine, 1667): Although this term
originated in the sphere of foreign trade, it achieved a wide currency not only in regions bordering on the Ukraine, but also in
the provinces of the Russian interior and even in Moscow itself. In
Russia the name "Cherkassk" was applied to Ukrainian *gorilka*,
which was prepared from wheat; unlike Russian vodka which
was prepared either entirely from rye, or from a mixture of rye,
oats and barley, or from rye with the addition of wheat.
"Cherkassk wine", or *gorilka*, was brought from the Zaporozhye
Sech, and received its name from the town of Cherkassy, close to
the then capital of the Ukraine, Chigirin; or more precisely from
the Cherkassk Cossacks. It was the Cherkassk Cossacks who in the
latter half of the sixteenth century began to bring their *gorilka*
through the Muscovite state for re-export to north-east Russia.
Cherkassk vodka was of low quality, sometimes very badly purified or not purified at all, and was worth considerably less than
Moscow vodka. A significant part of it was therefore bought up by
enterprising Russian merchants to be sold in the taverns of
Russia, while Russian vodka was exported or kept for the tables of
the ruling classes. The reasons for the poor quality of Cherkassk
gorilka lay in the technology of production: the makers did not

employ the extensive series of purifying processes used in Russia.[8]
The different raw materials also had an effect on the taste. In
Russian usage the term "Cherkassk wine" persisted until the
Crimean war of 1853–6, after which supplies of Ukrainian vodka
to the markets of the central provinces of Russia were cut off.
From 1861 Ukrainian vodka was made from potatoes.

Orzhanoe vintso, zhitnoe vino (rye wine): this term reflects the
fact that until the middle decades of the eighteenth century, that
is, for three centuries, grain spirit in Russia was distilled exclusively from rye, and for another century after that, rye continued
to constitute on average no less than half the base for Russian
vodka. Only from the middle of the nineteenth century did Russia
make a gradual shift to the use of wheat and potatoes for the
production of vodka. However, the use of a grain base remained
predominant until 1917.

Zel'eno-vino (herb wine), *zeleno-vino* (green wine), *khmel'noe vino*
(hop wine), *zel'e pagubnoe* (fatal herb): this group of terms is
encountered over three or four centuries, mainly in the everyday
language, in artistic literature and in folklore. However, these
terms have a real basis and signify any kind of grain spirit, of any
quality, that was flavoured with spices, or with aromatic or bitter
herbs, usually hops, wormwood, anise, pepper or some combination of them. The correct term for vodka of this category is *zel'eno-
vino*, from the Old Slavonic *zel'e* (grass, herb). It was distilled
together with herbs. At times such vodka could have a slight
greenish tinge, but the good quality product was colourless. It is
therefore incorrect to trace this term back to the word *zelen'*
(greenery) or to understand it as meaning "green wine".
Nevertheless, this error is very often made.

Gor'koe vino (bitter wine): this term, first recorded in 1548 and
later appearing afresh in 1787, achieved a particularly broad distribution in the second half of the nineteenth century and at the
beginning of the twentieth as a synonym for vodka. Its meaning
is close to that of the previous group; the term denotes vodka distilled with bitter herbs, wormwood, birch and oak buds, among
other things. However, this original meaning was lost at the end
of the nineteenth century and the term was completely reinterpreted, taking on the figurative sense of the word "bitter" – that
is, a drink bringing a bitter, unhappy life. The phrase "to drink

bitterly" came to mean to drink long and desperately. Therefore the term is now used only in literary contexts, as a metonym, and not as a normal designation for vodka.

Figurative and Slang Terms for Grain Spirit in the Eighteenth and Nineteenth Centuries

The eighteenth and nineteenth centuries saw the rise of various usages replacing the official or habitual terms for "grain wine". Another such term, "green wine", had been inherited from the old, pre-Petrine Russia; in the eighteenth and nineteenth centuries this term acquired a euphemistic force which it had not possessed earlier.

Euphemisms for words considered unseemly were characteristic of the nineteenth century, in which the Russian middle class developed. Metonyms, replacing a word with a related one, were more characteristic of the eighteenth century with its aspiration to cultivated classicism. Slang expressions for vodka became current in the Russian language mainly from the middle of the nineteenth century, and in any case only after the Patriotic War of 1812. The majority of substitutes for the basic, official, "normal" terms for grain spirit arose with the early development of the capitalist system in the eighteenth century.

The rise of these new forms reflected the loss of the standard quality of "grain wine", owing to changes in government policy on distilling. The exclusive state monopoly was abolished, and the right to engage in distilling was extended to the gentry as a whole, as a step in the introduction of numerous privileges, incentives, restrictions and bans for various social categories in Russia. In the distilling industry this created a chaotic situation which led to the appearance of grain spirit of the most diverse quality, as will be described later.

Official state policy on distilling did not change between 1649 and 1716, at least as far as the law was concerned. But between 1716 and 1765 a total of fifteen new decrees were issued on distilling and the production of grain spirit, and between 1765 and 1785 thirty-seven decrees. The final decree "On the Granting to the Gentry in Perpetuity of Permission to Engage in Distilling", which saw the state move out of producing vodka and abandon attempts to control its production by private entrepreneurs, was promulgated on 16 February 1786. This had the effect of consummating, through a firm agreement, the process of decentralization of vodka production which had continued throughout the century.

The first steps had been undertaken in the reign of Peter I, as a means of lessening social tensions within the country. In the event, the move brought great harm to Russia and to the Russian people, who through euphemisms and slang expressed their opposition to this "vodka freedom" through which "enlightened monarchs" tried to ensure internal order in Russia. Language was especially bitter after such social explosions as the Bulavin uprising of 1708 and the peasant war of Yemelyan Pugachev in 1774 and 1775. These terms are not only of historical and social significance, but are also important for our research because they help explain the emergence and consolidation of the term *vodka* as a substitute for *vino*. It was during the eighteenth century that people increasingly ceased to call vodka *vino*; the drink was not yet called *vodka* in official documents, but the official term *vino* (which was also, of course, a pure euphemism) came more and more to be replaced by jargon synonyms – which were, moreover, of a mainly disrespectful character. Thus in the early decades of the eighteenth century we note the appearance of such terms for vodka as the following.

Tsar's madeira: the irony of this euphemism was abundantly clear to the people of Peter the Great's time. It refers to the "assemblies" organized by Peter I, in which nobles and merchants had equal rights to participate and where any who had offended the Tsar were sentenced to drink huge cups of fortified wines – madeira, sherry or port – or spirits, which quickly rendered them staggering drunk. For the common people the vodka of Peter's time, with its deteriorating quality, was the "madeira" which they received free of charge, on the Tsar's account. In St Petersburg one cup per day was handed out free to all builders, road workers, stevedores, shipyard workers, soldiers and sailors. Thus the term meant vodka for the masses, prepared in the Tsar's taverns.

French of the Fourteenth Class: this name for vodka was widely current among state officials not only in the eighteenth century but to an even greater degree in the nineteenth; although it arose at the beginning of the eighteenth century as an ironic comment on the Petrine table of ranks. The term directly reflects the low quality of vodka in the Tsar's taverns. Vodka at this time was commonly diluted with water infused with tobacco.

Petrovskaia vodka (Peter's vodka): this term contains both a precise reference to the period when it originated and a comment on the quality of the grain spirit of the time. From the name Peter we know that this is not the former high-quality spirit but a new product of the Petrine era. From *vodka* (or *vodichka*, or *vodchonka*), we know that it is not worthy of the name "grain wine"; it is just "Peter's water", or not even water, but *vodka* – a contemptuous diminutive.

Today the term is incorrectly used on labels for so-called "old vodkas". Originally, far from indicating antiquity, it meant a new product, and an inferior one in comparison to the vodkas of the sixteenth and seventeenth centuries. The word *petrovskaia* has, absurdly, been taken to suggest that vodka of this type was the favourite of Peter I. As this Tsar often stated, his favourite was anise vodka; that is, a triple-distilled vodka diluted with aniseed water and distilled again.

Ogon' da voda (fire and water): this is a typical euphemism from the second half of the eighteenth century. It was used in a favourable sense by the lesser nobility and the provincial gentry, and among state functionaries, to describe high-quality, purified varieties of privately produced vodka. One might recommend that it be revived now for a good brand of vodka. The name suggests a picturesque reference to the two constituent parts of vodka: spirit and water.

Khlebnaia sleza (tear of the grain): this term, a euphemism with clearly positive overtones, was current mainly during the second quarter of the nineteenth century. It was used to describe privately produced vodkas of exceptionally high quality. In the second half of the nineteenth century the phrase, which by this time had lost its original significance, was used in the middle-class, petty official milieu as a euphemism for any vodka. It might be recommended for revival as the name of a vodka of the very highest quality, produced by the most exacting techniques or from especially high quality ingredients.

Sivak, sivukha: these are slang terms from around the end of the eighteenth century. They are masculine and feminine variants, reflecting uncertainty about the word to which they refer, which could have been either *vodka* (feminine) or *vino* (neuter). The words are derived from the adjective *sivyi*, used to describe greying hair. They were usually applied to vodka of extremely low quality or to vodka that was completely unpurified, and hence cloudy or greyish in colour.

Polugar, peregar: these are variants of a slang term from the middle of the nineteenth century, referring to a thin, low-quality grain spirit consisting of the last fraction of distillation and possessing a "burnt" smell or some other strong, unpleasant odour. The words originated from a corruption of a technical term for grain spirit, and persisted into the modern period as vague terms for any vodka of low quality.

Brandakhlyst (or more precisely, *brandakhlest*): another slang term from the mid-to-late nineteenth century, derived from the German word for spirit, *Branntwein*. The term was used mainly for poor-quality potato vodka imported at low prices from the western provinces of the Russian Empire – Poland and the Baltic region, and to some extent from the Ukraine and Belorussia (Volynia). By the end of the nineteenth century the terminological significance of the name had been totally lost, and it was applied to any low-quality vodka, irrespective of where it was produced. Nevertheless, the term *brandakhlyst* was still used mainly for cheap potato vodka, which until 1896 was available in the central regions of Russia. The term indicates clearly the unfortunate consequences of drinking "German" (in fact, mainly Polish) vodka – the word *khlyst* is a corruption of the verb *khlestat'*, in the sense of "to vomit" or "to induce vomiting".

Samogon, samogonka: these words appeared after 1896 and referred to home-distilled vodka, preparation of which had been outlawed since the original introduction of the state vodka monopoly. In general, the term *samogon* signified unpurified grain spirit of a quality immeasurably lower than that of the state-produced monopoly vodka. It should be noted that this was a slang term which arose in an uncultured, illiterate lower-class milieu. Until the beginning of the twentieth century the word *samogon*, in the sense indicated above, did not exist in the Russian language. Instead it had two specialized meanings. In the hunting terminology of Siberia, *samogon* referred to hunting on skis, as distinct from hunting (*gon*) on horseback and with dogs. In the jargon of distillers, *samogon* referred to the best fraction of spirit, from which the highest-quality, most refined liquors were prepared.

As can be seen, the term *samogon* with reference to vodka could not be borrowed either from the terminology of distillers or from the language of Siberian hunters. It appeared in the southern and south-eastern provinces of Russia, between the Ukraine and the Volga, that is, in provinces with a mixed non-Russian or partly Russian population, where the feeling for the language, for the purity of its definitions and meanings, was always weaker.

Industrial and Technical Terms for Grain Spirit and Its Quality

Raka: this name was given to the product obtained from the first stage in distillation of grain spirit, directly from the mash in the vat. The process continued to the point where the liquid in the mash disappeared and the solids began to burn. *Raka*, as a technical term, appeared simultaneously with the production of vodka, or more exactly, with the technology of distilling. In medieval sources the word is sometimes equated with "stinking vodka". The use of this phrase can be considered as one of the first signs pointing to the eventual designation of the product as a whole by the word "vodka".

The term *raka* has its origins in the Arabic word 'araq (compare Turkish *raki*, Indonesian *arak*), which was used for date vodka in Asia Minor. The term is encountered in Byzantine sources dating from as early as the third century, and is also found later, from the eighth to the tenth century. It was borrowed by the Russians from Byzantium, either directly or by way of Bulgaria and Moldavia, where any fruit vodka, the product of distilling fruit or berry wine or a mash of fruit and berries, was known as *rakiia*.

In Asia Minor, Greece, Bulgaria, Wallachia, Moldavia and Serbia – and among the Eastern Slavs, in Belorussia[9] – this term signified a finished product of high quality. It is therefore curious that in Russia it should have been used to denote the crudest type of intermediate product. This paradox is explained by the fact that the first people to translate the term were monks. In the Greek text of the New Testament the Aramaic word *raka* or *rakha* is used to mean empty or foolish, and therefore was interpreted in a correspondingly negative manner, irrespective of the context, by Russian translators. Encountering difficulty in finding a Russian equivalent for the word *raka*, the monks interpreted it in a figurative sense as referring to someone who drank *raka* – a drunkard, and consequently a fool. In a church dictionary published at the end of the eighteenth century the word *raka* is taken to indicate a brainless person, unreasoning and contemptible. A whole set of negative definitions is provided for this term, referring not simply to a drunken person, but to an exceedingly drunken one. Probably this corresponded to the difference between drunkenness as induced by the old drinks and by the new spirits.

Russian distilling lacked its own terminology, since it was based on Byzantine techniques (perhaps Greek monks fleeing from Constantinople in 1453 brought distilling apparatus with them, or perhaps such apparatus was brought in with the baggage of Sophia Palaeologus in 1472). It is easy to see why the Russian distillers would have employed the biblical term *raka* in its Russian sense, which had become familiar after five centuries, to denote the first, crudest product obtained in distilling "grain wine". Merely sampling this would have been enough to induce severe drunkenness and an appalling hangover. It was entirely logical for the distillers to describe this crude intermediate product as *raka*, implicitly defining it as dangerous, worthless, and not yet fit for use.

The most interesting conclusion that can be drawn here is, however, that the use of biblical terminology to describe such an "ungodly" pursuit as alcohol distilling indicates specifically and convincingly that in Russia, as in Western Europe, distilling arose in the monasteries. But while in Catholic Europe this industry remained for a prolonged period in the hands of the Church, in Russia it quickly became wholly or almost wholly concentrated in the hands of the secular power – the monarch, the state administration, and the treasury – in the form of an official monopoly. This occurred before the end of the fifteenth century, and was consolidated early in the sixteenth century under Ivan IV.

Early in the seventeenth century the Church also lost its privilege of being allowed to produce vodka for its own internal purposes, which illustrates the determination of the secular power not to share with the Church this powerful and dangerous source of financial and social influence. In the eighteenth century the Church suffered a similar blow when the right to engage in distilling was granted exclusively to the gentry; the copper still apparatus, such as pipes and vats, was taken from the monasteries under the pretext of using it for military purposes. In this way, "vodka was taken away from the Church", and this led to one of the distinguishing features of Russian historical development. The event fixed the generally negative attitude of the Church to the "Devil's poison", and brought about a clear division of influence between Church and state; the first was permitted only to concern itself with people's souls while the second assumed the right to influence their bodies. The sole reminder of the Church's participation in establishing grain spirit production in Russia was the word *raka*; and that only until the nineteenth century.

During the second half of the nineteenth century the term *raka* began to be replaced by the modern, purely technical term "first distillation". This change in terminology signalled the complete transfer of distilling to the hands of the Russian industrial bourgeoisie, and the final break with medieval patriarchal traditions in grain spirit production.

Prostoe vino (simple wine): This term, introduced officially in the Statutes of 1649, referred to grain spirit produced by single distillation from *raka*; that is, double distillation of the alcohol brewed in the original mash. "Simple wine" was the main intermediate product used to create the usual, widely distributed type of "grain wine", known as *polugar*, and also all the other official types of vodka and grain spirit in Russia.

Polugar: in the distillers' jargon this term denoted not the spirit obtained through distilling *raka*, but "simple wine" diluted in the proportion of one quarter with pure cold water (one "bucket" of water was added to three "buckets" of "simple wine" — the term "bucket" is defined later in this chapter). The name was inspired by the fact that the mixing process was usually followed by a test of the quality of the product. A special metal cylinder with a volume of about half a litre and with a dividing line marking half the volume was filled with spirit, which was set alight. If the spirit burned down to the line — that is, halfway — it was considered fit for use. The simplicity of this test had vital significance in maintaining a single standard of quality for vodka throughout the state, since every drinker had the opportunity to conduct such a test of the liquor he had bought. This effectively stamped out any meddling with *polugar*, and as a result this relatively weak variety of "grain wine" achieved a particularly wide distribution in the sixteenth and seventeenth centuries.

The popularity of *polugar*, and especially of the simple technical test associated with it, influenced the everyday terminology of grain spirit that we noted earlier, serving particularly to reinforce such names for vodka as "burning wine". In the Ukraine, where for centuries *polugar* remained the principal and even sole form of "grain wine", this later resulted in the name *gorilka* becoming established even for vodka of high quality. In Russia, vodka production achieved progressively higher levels of technical development from the late seventeenth century, and especially in the early eighteenth, as distillers aimed at more concentrated and better purified forms of the drink.

During the nineteenth century the word *polugar* was transformed from a technical term into slang implying any form of non-standard grain wine of low quality. Beginning with the eighteenth century not only "simple wine", but also *polugar*, came to be viewed by more discriminating drinkers not as a finished product but only as an intermediate one, even if it was already fit for use. The official strength of *polugar* in the middle of the eighteenth century did not usually exceed 23 or 25 per cent of alcohol by weight, but since this type of vodka did not undergo special filtration its taste and smell were unpleasant because it was contaminated by congeners such as "fusel oil" – mainly butyl and iso-amyl alcohol. These contaminants were generally removed in a relatively easy if primitive fashion by passing the *polugar* through a wire-gauze box filled with birch charcoal made from twigs the thickness of a pencil. *Polugar* "with twigs" had been purified, but "without twigs" it remained of low quality. Distorted recollections of this have remained in Russian slang to the present time. Any low-quality vodka, inadequately distilled and filtered, is now described in vulgar terms as a "twig", although logically it should be called the precise opposite, "without a twig", that is, unfiltered and unrefined. However, *polugar*, not to mention "simple wine", was not filtered if it was produced in the usual fashion. Only the superior varieties of grain spirit were favoured with this treatment.

Pennoe vino, *pennik*: these names denoted the best variety of vodka that was obtained from "simple wine". The name was not derived from the word *pena* (froth), as is now often assumed, but from *penka*, which in the seventeenth and eighteenth centuries had the sense of the best, the essence, of any liquid. (As an analogy, we may take the word "cream" in its figurative sense, as in "the cream of society".) In Old Russian the word *penka* meant in addition the topmost layer of any liquid. In vodka production, the term was applied to the best fraction produced during distillation. No more than a quarter or even a fifth of the volume of *raka* went to form this "cream", which was, moreover, obtained though very slow distillation. Similarly in the next stage, it was the lightest fraction of "simple wine", obtained in the same fashion and possessing the highest alcohol content, that went to produce *pennik*. From the middle of the nineteenth century until the twentieth, this fraction, instead of being called "cream", came to be called the *pervak*, that is, the first, and from 1902 the term *pervach* was officially applied to it. A hundred buckets of *pervach* diluted with twenty-four buckets of pure soft spring water created *pennik* or *pennoe vino*. Contemporaries who described *pennik* stressed not so much its strength as the fact that it was "good spirits", pure, soft, and pleasant to drink. *Pennik* always underwent filtration through charcoal, though it had less need of this than other varieties of vodka.

As well as *pennik* there were other degrees in which "simple wine" was diluted with water, giving weaker and cheaper varieties of vodka. These were intended for various categories of consumers, on the basis both of ability to pay, and of age and sex.

Trekhprobnoe vino (triple-tested wine): this term denoted vodka obtained by diluting a hundred buckets of spirit with thirty-three – but in unscrupulous hands up to seventy-three – buckets of water. In this it resembled *polugar*, but it differed from *polugar* in its lesser strength and poorer quality, since it was made from the "simple wine" left over after the *pervach* had been distilled off, while *polugar* was prepared from untouched "simple wine". The production of this type of vodka reached a high point during the first half of the nineteenth century, and "triple-tested wine" was the usual vodka on sale in taverns for the common people, the *muzhiki*.

Chetyrekhprobnoe vino (quadruple-tested wine): this term referred to spirit which had been watered down even further and was therefore cheaper still. To a hundred buckets of "simple wine" fifty buckets of water were added.

Dvukhprobnoe vino (double-tested wine): this term – illogically in view of the above – denoted vodka obtained by diluting a hundred buckets of the original distillate (from which the *pervach* had not been separated off) with a hundred buckets of water. Among the common people such a drink was considered "women's wine", and was one of the usual offerings of taverns in the first half of the nineteenth century.

Dvoinoe vino, *dvoennoev vino*, *peredvoennoe vino* (double wine): these terms were applied at various times to grain spirit obtained by the distillation of "simple wine"; that is, the resulting spirit which was the product of triple distillation, from *raka* to "simple wine" to "double wine".

From the eighteenth century this stage was considered obligatory for good distilling aimed at obtaining a high-quality product. In the distilling carried on by the gentry in the eighteenth century, and especially during its second half, the standard intermediate product was considered not to be "simple wine" but "double wine". It was spirit of this type that served as the basis for all the experiments aimed at perfecting Russian vodka during the eighteenth century, and "doubling" was officially recognized from 1751 as the normal and even elementary guarantee of quality. At that time the strength of "double wine" from various distilleries was around 37 to 45 per cent alcohol by weight. Very often, in around half of all cases, "doubling" involved distillation with added spices and aromatic ingredients, or with the addition of diverse absorbents and coagulants. More will be said on this later, in Chapter 4.

"Doubling" was thus a complex operation which not only increased the concentration of spirit, but also improved the general quality of the drink by cleansing it of contaminants and unwanted flavours.

The idea of diluting "double wine" with water was a natural one, since the gentry and monks, who had been brought up on grape wines, were unaccustomed to such strong spirits. In the nineteenth century, as we saw earlier, "simple wine" was also as a rule diluted with water, not so much because of its alcoholic strength as for tradition's sake, and because it was noted as a characteristic of diluted spirits that they were relatively free of contaminants, and in the first instance of fusel oil. At least where treatment of the already distilled product was concerned, Russian vodka production thus achieved a high technical level long before the discoveries of scientific chemistry were brought to bear on it. This refinement was the result of historical experience, of combining tradition with experiment.

Vino s makhom (wine with a wave): as we saw earlier, the practice of diluting grain spirit of the same concentration with varying quantities of water or, on the other hand, of mixing various fractions and distillates of grain spirit with water, was a basic principle of Russian vodka production. Various types of "grain wine" were created by varying the proportion of spirit. But alongside this predominant idea in Russian vodka production, the idea had already arisen around the end of the seventeenth century of using a mixture of distillates to increase the alcohol content of various drinks, and thus creating other varieties of "grain wine", different from vodka.

These varieties were referred to as "wine with a wave" – that is, a dismissive gesture – a name which was part distillers' jargon and part everyday usage. The most usual form which this drink took was a mixture of two-thirds "simple wine" with one-third "double wine". Other proportions were used as well, with "double wine" making up from a quarter to a half of the mixture. However, these mixtures of spirits, which were not diluted with water and which consequently were not true vodka, did not become popular in Russia. Their drawbacks included first of all a high price; but more importantly they had an effect more devastating than that of pure "double wine", though technically they were weaker.

The reasons for this will be explained later, in Chapter 4. Here we shall confine ourselves to noting that in the early decades of the eighteenth century "wine with a wave" was already viewed as an aberration and as a departure from the normal course of vodka production. Such mixtures were accepted as technical curiosities, but not as drinks, since it was noted early on that they were virtually impossible to purify, even with the best filters. At the same time, Russian vodka distillers did not draw back from producing a drink with a higher alcoholic content than "simple" or "double" grain spirit, and therefore set off along the path of creating a more and more refined, concentrated spirit, which had subsequently to be diluted with water. This path came to represent the basic thrust of Russian vodka production as early as the 1720s and 1730s. From this time on "wine with a wave" was not only outlawed by official legislation, but was considered by the private producers themselves to be in the same category as *raka*.

Troinoe or troennoe vino (triple wine): this name was given to the distillate prepared from "double wine"; in fact "triple wine" was quadruple-distilled spirit. It began to be supplied by private distillers to Russian connoisseurs during the second quarter of the eighteenth century, and its production was made possible by the perfecting of the distilling apparatus. "Triple wine" was used for the production of highly refined vodkas for domestic consumption; in practice, aromatic herbs were always added to it.

During the nineteenth century "triple wine" was used as the basis for the best factory-produced vodkas of the pre-monopoly period. The vodkas of the Popov factory, for example, were known to consumers by the name "popovka" or "triple popovka". "Triple wine" in its undiluted form had a strength of approximately 70 per cent alcohol. The great chemist Dmitry Ivanovich Mendeleyev regarded it as the classic basis for the preparation of vodka through the addition of water.

Chetvernoe or *chetverennoe vino* (quadruple wine): this was often called *chetvernoi spirt* (quadruple spirit), and in the eighteenth and early nineteenth centuries was also known by the Latin name *vinum rectificatissimum* (highly rectified spirit). This product of quintuple distillation was already known at the end of the seventeenth century, in 1696, when it was distilled in the Tsar's palace laboratory; but it was prepared rather infrequently, and only for scientific and medicinal purposes. The strength of this distillate in the late nineteenth century was around 80 to 82 per cent. In the eighteenth century, however, Academician Tobias Lovits had obtained so-called "waterless spirit" – absolute alcohol, almost 96 per cent pure.

On the basis of "quadruple spirit", herbal infusions were prepared during the 1760s without the addition of water. These were classed not as vodkas, but in a special category of alcoholic liquors.

A number of vodkas were also created on the basis of "quadruple spirit". There were two groups of these: aromatic, and sweetened (the so-called ratafias). The latter were also coloured with berry syrups or with other plant extracts. Ratafias, like herbal spirits, did not become widely distributed. They had already become a rarity in the nineteenth century. The basic reason for this was their high cost and the unprofitability of producing them, since they needed much labour, some of it highly skilled, and the use of special materials such as isinglass.

Such highly refined alcoholic drinks could be produced and developed, at least on a small scale, in the private industries of gentlemen distillers at the height of the era of serfdom. The development of capitalist production, with its implacable demand that commodities should be profitable, led progressively to the disappearance of these drinks. The ratafias of the nineteenth century, and especially of its second half, were a counterfeit product completely lacking the exceptionally high quality attained during the eighteenth century.

The Historical Significance of Terms for Grain Spirit

From our survey of the terms associated with grain spirit in Russia between the fifteenth and nineteenth centuries, we can draw certain conclusions.

One is struck by the enormous multiplicity and variety of terms – technical, commercial and, especially, customary and everyday. In the first instance this reflects the diversity of quality and type of the various "brands" of vodka, although formally speaking the division of vodka into genuine brands – that is, varieties with their own names – began only in the era of modern capitalism. This variety of terms testifies to the fact that the development of vodka over a number of centuries was a constantly unfolding process whose general tendency was an often unconscious, but sometimes deliberate, striving to obtain the ideal, perfect product.

Detailing the content of the terms associated with vodka shows clearly that the central course of development in Russia was toward the creation of an alcoholic drink of the vodka type; that is, a drink obtained by diluting spirit with water. From this emerge the reasons why the Russian type of grain spirit drink ultimately received the name "vodka", even though this was not its original name. Moreover, it becomes clear that such a name for the drink could not arise immediately, since this name reflects the characteristic features, properties, and principle of composition of the drink, all of which crystallized only as a result of prolonged development.

At the same time it is logical to expect that, as an intermittently, often accidentally used name, "vodka" should have appeared at an extremely early stage in the production of Russian grain spirit, even when contemporaries were primarily conscious of some other, more characteristic feature of the drink – for example, its "burning" properties.

Analysis of the terms shows that in the early period when grain spirit was just beginning to be produced, Russian consumers were struck by the most diverse aspects of the quality or of the external features of this product (place of origin, physical characteristics, "burning" properties and so on). But for the people who were directly involved in producing vodka, the key characteristic of the drink was the fact that every stage in distillation required diluting the distillate with water, at times in large quantities, before a product was obtained that could be drunk by the consumer.

This tendency was characteristic only of Russian production, and arose under the influence of the Byzantine tradition of diluting any alcoholic drink with water before it was consumed. Thus, for example, the Russians not only diluted Greek and Italian grape wine, but also added water to mead.

This traditional practice of dilution with water came inevitably to be applied, even more thoroughly, to the new alcoholic drink "grain wine". With its characteristic taste and odour, "grain wine" was perceived as requiring dilution, especially since Russians were accustomed to pleasant-tasting, aromatic drinks based on such delectable ingredients as honey, aromatic herbs, berry juices and malt.

One can suggest with a good deal of confidence that originally in the fifteenth century, and even until the seventeenth century, the grain spirit that was distilled was diluted not simply with water, but with honey water; that is, water that had been lent an aroma through the addition of a small quantity of honey. In the *Domostroi*, a housekeeping text of the mid-sixteenth century, there is a definite statement that one should dilute "simple wine" with honey water rather than with plain water.

This explains in one way why the name "vodka" did not become widely current for centuries after the drink had begun to be produced. People did not consider that they were diluting grain spirit with water, but that they were infusing it with a weak solution of honey. On the other hand this shows no less convincingly that the name "vodka" did not drift by chance into people's ken in the eighteenth century. It was just at this time that "grain wine" ceased to be diluted with a honey solution, as people came to prefer to lend aroma to the drink and to banish unpleasant odours by adding herbs and spices. Moreover, by this time the dilution of grain spirit had come to be dictated less by Byzantine traditions than by technical considerations and by the historically formed habits of the consumer. "Wine with a wave" – that is, undiluted blended spirits – did not become established and did not enter widely into production and trade. The only drink with a future proved to be vodka; that is, spirit of any strength or any degree of refinement, but diluted with water after distillation.

Vodka therefore constitutes a specifically Russian form of alcoholic drink produced from grain spirit in a fashion determined by tradition and by historical forces, in line with the history of the technical development of Russian distilling. In other words, vodka arose in Russia, and in Russia no other spirituous liquor could have succeeded other than vodka.

In other countries of Europe – France, Italy, Britain, Germany, and Poland – the process of developing strong alcoholic liquors took another path: that of progressively increasing the strength through perfecting the distilling apparatus and increasing the number of distillations. The bizarre idea of using water to dilute a high-quality, concentrated product obtained through the careful use of advanced technology would never have occurred to the developers of cognac. It would have meant throwing away the results of all their exertions. Nothing would have suggested this

course to them, neither tradition, nor logic, and least of all technical progress. In Russia, on the other hand, everything conduced
to it: the demands of the Church; tradition; the taste for "soft",
pleasant beverages; the low quality of the product, which needed
to be improved; and the technical imperfections of the apparatus,
which forced the makers not simply to distil their product but also
to filter it.

Vodka arose in accordance with historical laws, and then in
just as historically legitimate a fashion opened up a path for itself,
vanquishing all the ancient national Russian drinks and ensuring
its hegemony by virtue of being the most acceptable and appropriate drink in the Russian historical context, with all Russia's drastic changes in economic and social conditions. To render these
conclusions still more concrete, we shall review one more group
of terms connected not with vodka as a drink, nor with its production, but with the measures of liquid volume in Russia. From
this it will be clear to what extent these measures were adapted to
and were dependent on such a commodity as vodka; and how the
changing character of liquid measurements literally allows one to
trace all the stages of vodka's displacement of other drinks, and in
the process makes it possible to fix more precisely the moment
when vodka became the unrivalled ruling beverage.

The History of Liquid Measurements in Russia

The oldest Russian unit for measuring liquids was the "bucket"
(*vedro*). This unit of volume was used from very early times
throughout the whole territory of Russia, in the various Russian
states; the different provinces used various versions which persisted for many years. The distinguishing feature of the bucket
was that it was the basic unit on which both larger and smaller
units were based. Large volumes were defined by the number of
buckets which made them up. The bucket was also divided into a
series of smaller units which were commonly used in daily life
and petty trade. In effect, the bucket as a unit of liquid measurement occupied the same position as the ruble later occupied in the
monetary system.

The first known mention of the bucket dates from the year 996
or 997. At this time its volume ranged from 12 to 14 litres in

different provinces. These units were used to calculate the volumes of the main alcoholic liquor of those times, mead. A vat (*varia*) of mead in various provinces consisted of 60, 63 or 64 buckets. No other measures are known from this period.

Later, in the twelfth century, we encounter such units of volume as the *varia pivnaia* (beer vat) and the *korchaga* (pot). The beer vat had a volume of some 110 to 112 buckets, and in the process of production 120 buckets of "thick" beer were taken from it, or 220–40 buckets of "thin" beer. This allows us to understand the term by which "thin" beer was sometimes known until the nineteenth century – *polpivo* (half-beer).

The first record of the *korchaga* dates from 1146. This was a measure of grape wine, a standard Greek-style amphora with a volume of 2 buckets or 25 litres. In the twelfth century, therefore, a bucket of grape wine had a volume of 12.5 litres. From the fifteenth century, when vodka first appeared, we encounter more precise indications of what kind of bucket the source had in mind: a bucket of mead, or a bucket of "church wine", and so on. In the sixteenth century we encounter for the first time the term *ukaznoe vedro* (official bucket). This signified that vodka had appeared, along with the monopoly on its production and sale.

The official bucket was equal to twelve "mugs" (*kruzhki*); a mug had a volume of 1.1 or in some places 1.2 litres. Simultaneously we find the term *vinnoe vedro* (spirit bucket), that is, a bucket clearly meant for vodka, equal to 12 mugs, and with a weight of 35 Russian pounds (about 14 kilograms) of pure water. In the seventeenth century this volume for the spirit bucket remained, but it was divided into 10 mugs, so that a seventeenth-century mug had a weight of $3^1/_2$ pounds of pure water.

From 1621 a new name for the spirit bucket appeared – "palace bucket" (*dvortsovoe vedro*). It was also known as the "drinking measure" (*piteinaia mera*) or "Moscow bucket". The "Moscow bucket" at this time was the smallest bucket in Russia, now equal to only 12 litres, while in other provinces the bucket still varied in volume from 12.5 to 14 litres. It is plain that the choice of this small volume for the Moscow bucket (which was, it should be remembered, a measure of vodka) was no accident. The Moscow bucket continued to be identified with the old, large bucket, and its contents sold for the same price. But purchasers received much less vodka from a bucket in Moscow than, for example, in Tver, where the bucket contained 14 litres. From this time the provincial buckets quickly disappeared. All the provinces introduced the Moscow bucket.

In this way, without the issuing of special orders or the application of pressure from the centre, the Moscow bucket spread rapidly throughout the whole state as the sole unit of liquid volume. Vodka quickly levelled out the provincial Russian units of measurement, which would have remained in use much longer if what had been involved had merely been grain, peas, milk, berries, apples or other such products. In the case of these goods Russian trading practice had developed the custom of measuring out dry and liquid volumes allowing a certain leeway, without bothering much about the odd extra mugful, especially in wholesale trade.

From the sixteenth century there were also precise divisions of the bucket into small measures. From 1531 the bucket was divided into 10 *stopy* (beakers) or 100 *charki* (cups). This division, as we shall see, was typical of vodka, although what was involved was not yet the Moscow drinking bucket, but the bucket in general. The cup contained 143.5 millilitres, or close enough to 150 millilitres, the Moscow norm for a single shot of vodka. (This led to the wide currency of expressions such as "to raise one's cup [*charka*]", "to drink a cup", "a cup of vodka", "he poured a cup for his guest, and drank two himself".) Only towards the beginning of the twentieth century was the volume of the cup changed somewhat, when it was reduced to 123 millilitres.

Another measure of the seventeenth and eighteenth centuries was the *kovsh* (scoop), of three cups or about 500 to 600 millilitres. Later, in the nineteenth century, this measure was transformed into the "bottle" or "vodka bottle" of 610 millilitres; in Tsarist Russia there also existed a "wine bottle" of 768 millilitres.

The eighteenth century saw the introduction in place of the beaker of a Western European measure, equivalent to a German *Stubbe* or English quart: the *shtof* of 1.23 litres. In line with this the cup changed its volume; the *shtof* was divided into 10 cups. The *shtof* in fact corresponded to the old Russian measure, the mug (*kruzhka*) of 1.2 litres, while a half-*shtof* was the equivalent of the usual bottle of 610 millilitres. However, in the process of becoming transformed from a commercial measurement to a domestic one, by the beginning of the nineteenth century the mug had come to have a volume of only some 750 to 800 millilitres, and was not taken up as a "Russian litre".

Apart from these main units of measurement, in the monasteries the *krasovul*, with a volume from 200 to 250 millilitres, continued to be used in various provinces to measure wine. The *krasovul* corresponded to the glass of those times and served as a church measure for grape wine, for which the cup was not used. It often happened that for lack of actual physical cups, which had been forbidden in the monastic way of life, the monks drank vodka by the *krasovul*. Another of the ancient Russian measures which was retained for many years, right up until the twentieth century, was the *chetvert'* (quarter), between 3 and 3.25 litres. This was a bottle containing a quarter of a bucket. In some provinces where the quarter had a rather greater volume of about 3.5 litres, it was called a "goose" (*gus'*). In order to distinguish it from the domestic bird, this measure was assigned the feminine gender.

From 1648 to 1701 vodka was sold exclusively by weight, and was not measured in units of volume. This change was brought about first of all because at this time a change of measuring units had just been introduced, and secondly with the aim of preventing any tampering with the composition of vodka. Special buckets which were sent out from St Petersburg to all Russian taverns were supposed to hold a uniform weight of 30 pounds (about 12 kilograms) of vodka on all the territories on the European side of the Urals and 40 pounds (about 16 kilograms) in Siberia; if the weight exceeded this figure, it meant that the vodka had been watered down. Through this ingenious ploy the authorities hoped to force the observance of the state standard for the quality of vodka. In practice, however, this method proved ineffective, since people were accustomed to being on their guard against attempts to short-weight them, but not against attempts to give them too much weight.

According to a decree of 1721 a soldier was entitled to an allowance of two mugs of vodka per day. This amounted to some 3¾ pounds of vodka or about 1.5 litres per day of a drink with a strength of about 15 to 18 per cent; that is, vodka diluted with water in the proportions of "triple-tested wine".

Meanwhile, large units of liquid volume were established by decree. All these were related to the measures used for vodka.

From 1720 a standard barrel, known as a *sorokovka*, contained 40 buckets. This barrel was used as a unit of measurement of grain and water, and only incidentally for wine and vodka. In 1744 a so-called "official barrel" (*ukaznaia bochka*) of 30 buckets was introduced specifically as a measure for vodka. Besides this a "vodka keg" was devised, in which the superior types of vodka were measured and stored. This keg had a volume of 5 buckets.

In addition there existed a "heavy spirit barrel" of 10 buckets, while here and there in the provinces "mead barrels" of 5 and 10 buckets had been retained. In Belorussia and in Smolensk and Bryansk provinces there was an 18 bucket "Smolensk barrel" until the end of the eighteenth century; this was also used as a measure for various alcoholic drinks including mead, beer and vodka. The largest provincial unit of measurement for vodka was the *sopets*, of 20 buckets.

At the beginning of the nineteenth century, and especially after the Patriotic War of 1812, three main measures of volume remained in use in the wholesale vodka trade. These were the official barrel of 30 buckets, the "heavy spirit barrel" of 10 buckets, and the "vodka keg" of 5 buckets. In retail trade the main units of measurement until 1896 were the bucket (for consumption off the premises) and the cup (to be drunk on the spot). Only toward the very end of the nineteenth century, from around 1885, did vodka begin to be packaged and sold in Russia in the manner to which we are accustomed today. The bottles were of 610 millilitres and 1.22 litres; today's 500 millilitre and 1 litre bottles appeared during the 1920s.

Thus from the fifteenth century measurements of liquid volume in Russia developed in an uninterrupted relationship to vodka, and were adapted to the standard units of measurement established for the production and individual consumption of vodka. From the sixteenth century all other measures of volume connected with other alcoholic drinks, such as grape wine and beer, rapidly died out, apart from the standard wine bottle.

Although the basis of the new vodka measurements remained the ancient, time-honoured Russian unit of volume, the bucket, its volume was reduced from 14 to 12 litres. From the sixteenth, and especially the seventeenth, century the bucket was divided into 10 and 100 parts, whereas a decimal system of units had barely been thought of elsewhere in Europe. The large unit measurements of liquid volume, barrels, were also arranged according to a system not far from the decimal one, in units of 5, 10, 20, 30 and 40 buckets.

All of this testifies indirectly to the fact that while the trade in vodka began later than the trade in other types of goods, it developed rapidly and very soon came to encompass the whole of Russia. This forced the adoption of completely new units of measurement, different from the former local units of the various principalities and provinces. The new system needed to be simple and readily understood by all, and to reduce inconvenient calculation; hence the use of the decimal system.

Also of interest are the links between individual units of measurement of vodka and the history of this drink. The original unit, the mug or beaker, close to the litre, was replaced in the eighteenth century by the much smaller cup. This shows that until the eighteenth century vodka was extremely weak, and was diluted with water immediately after the distillation of the *raka*, that is, after the first degree of refining. From the eighteenth century vodka was prepared mainly from "double" and even "triple" spirit, and its strength increased significantly.

In the course of the nineteenth century, and especially following the reintroduction of the monopoly in 1894, the strength of vodka became established at modern levels. For over-the-counter retail sales the cup of 145 millilitres was supplemented with a smaller unit of volume, the half-cup of 72.5 millilitres (or in practice, 70).

Thus the units in which vodka was sold for consumption on the spot showed a tendency to decline in direct proportion as the strength of vodka increased. This suggests that the scale of vodka consumption in Russia remained relatively stable over a very long period, if we consider the volumes consumed in the light of their changing strength. The hypothesis is confirmed by statistics from the last decades of the nineteenth and the early years of the twentieth century, when for the first time enough detail was given to make accurate calculations. Consequently, the widespread view that the growth of drunkenness in Russia was caused by the expansion in the production and use of vodka is incorrect – or at least, much more open to dispute than it might seem.

Throughout its development from the fifteenth to the nineteenth century vodka underwent constant changes. The main question is precisely how it changed. To come up with an answer which is historically precise and technically competent, we need to make a detailed examination of the technology of production of vodka from the fifteenth to the nineteenth century. This is indispensable for two reasons.

First, we need to see how the production of vodka evolved – whether this was along the path of improving its quality, or along that of reducing the cost of production without improving quality (or even reducing it). In the history of industrially processed foodstuffs there are many examples of economic necessity and the laws of capitalism forcing development along the latter path. But there are also products which escape such a fate. Therefore this question demands special study.

Second, we know very little about the phases of development through which the technology of vodka production passed. For the present we cannot pinpoint the apogee of development of vodka or even say whether such a high point existed. Nor do we know whether the development of vodka proceeded on a steady upward path towards improvement, or whether there were dips in the graph of quality.

All this still needs to be explored. For this purpose it will be necessary to study the chronological succession, establishing how vodka was produced and how these techniques changed in each period. But before we take up the history of the technology of vodka production in the next chapter, we need to dwell in more detail on the term "vodka". This term was not examined along with others earlier in this chapter, since it emerged in its own right as an official state and national category only with the reintroduction of the state monopoly on vodka at the end of the nineteenth century. Despite this, the word "vodka", embodying the concept of Russia's national variety of grain spirit, arose in the middle of the eighteenth century, and in fact was used in other senses much earlier.

The foregoing investigations have brought us to the point where we can explain the concept of vodka and the essential character of Russian grain spirit. So it is now appropriate to examine the word "vodka" and its true meaning.

The Rise and Development of the Term "Vodka" from the Sixteenth to the Twentieth Century

In our review earlier in this chapter of the significance of the term *khlebnoe vino* (grain wine), we did not observe at any point the use of the term "vodka". Nor did we comment historically on the etymology of the word, if we except a short linguistic commentary in Chapter 1.

This is due above all to the fact that "vodka", as an established official term employed in state legal documents, came into use very late. The first recorded use of the term is in the decree of Elizabeth I on "Who is to be Permitted to Possess Vats for the Distillation of Vodkas", issued on 6 June 1751. The term next officially appeared almost 150 years later, around 1900, in connection with the reintroduction of the state monopoly on the production and sale of vodka.

In popular speech, however, the word "vodka" arose relatively early and was employed for several centuries as a slang term. Its broad spread in this capacity took place mainly during the reign of Catherine the Great, for reasons which will be discussed below.

The word was known considerably earlier than the middle of the eighteenth century, but its meaning both before and during that century did not coincide with the present one – that is, with the meaning which was placed on it at the beginning of the twentieth century. For this reason, and also because the word "vodka" was perceived very much as slang, the term is practically impossible to find in any authoritative dictionary of the Russian language, or in reputable literature, right until the 1860s.

The instances of the use of the word which have come down to us in various sources, including literary and documentary texts, are in large measure casual and do not give a full idea of how it developed in all its meanings. Nor do they give a real idea of when this word first appeared as a term for an alcoholic beverage, or how the concept developed.

Nevertheless, we shall try to assemble and set in chronological order all the available information bearing on the time when the word "vodka" first appeared and on the various senses in which it was used. This will help us understand how, when and why a word, which until the thirteenth century had simply meant water, altered its meaning and came to be applied to the Russian national drink.

There is no surviving text from the fifteenth century which mentions the word "vodka" in a sense close to that of an alcoholic substance. In the Novgorod chronicle for the year 1533 the word is mentioned as the name of a medicine: "Take *vodki*, apply to the wound and press out." "For my headache, the master ordered me to take . . . *vodki* from his medicine chest." From these texts it is clear that what was understood by "vodka" in the sixteenth century was not a decoction made by boiling herbs in water, but a tincture, that is, an infusion in alcohol. A tincture would have been an effective disinfectant for a wound, and only an infusion in spirit could have been stored for long periods in a medicine chest. Decoctions, which did not keep, were prepared on the physician's instructions, at home and for immediate use.

Thus from an early date infusions in alcohol were described as "vodka". But what was the reason for this term being transferred to an alcoholic drink? For this it was essential that water (*voda*) should somehow have been involved in the formulation of the medicine. Otherwise it is impossible to explain how the name of an infusion in alcohol could have become linked with the word "vodka".

In fact, Russian medicine of that time also used such terms as *vino martsial'noe* (martial wine) and *vino khinnoe* (quinine wine), together with *vodka finikolevaia* (date vodka). Let us compare the contents of these preparations. We know that medicines bearing the name "wine" consisted of infusions in grape wine, either white or red. Thus, *vino martsialnoe* was an infusion of iron filings in red burgundy, while *vino khinnoe* was an infusion of quinine bark, usually in a German white wine.

The name "wine" for these medicines is thus fully justified, and reflects the real content of the preparations, which were used as tonics. We know from the recipes that infusions of medicinal herbs which were described as "vodka" rather than "wine" were either infusions in distilled alcohol which were added to water immediately before use (for example, a drop of the infusion to a spoonful of water or a spoonful of the infusion to a glass of water), and which therefore tended to be seen by the patients as water-based rather than alcohol-based; or else were the product of redistilling "simple" grain spirit with medicinal herbs, in which case the resulting extract was then mixed in a proportion of one to three with boiling water. Thus, for example, "white vodka for strengthening the stomach" was prepared by mixing various spices (sage, anise, mint, ginger and amaranthus) with a combined weight of about half a kilogram. This mixture was then infused in five litres of "simple wine". This infusion was redistilled, then diluted with boiling water in the proportion of one part of water to two parts of spirit. From the very first this "medicine" received the name *vodka*, in line with the method of its production and the linguistic usage of the time.

Thus the technology of medical or, more precisely, pharmacological practice gave rise to a product which had little in common with tinctures as understood in the Western European pharmacology of that time, and which consequently had to be given a new and appropriate name – "water-diluted tinctures", or "vodkas". At the same time as apothecaries in Western Europe were striving to concentrate their medicinal preparations to the greatest possible degree, to produce remedies that were small in volume and hence portable, their counterparts in Russia in the sixteenth and seventeenth centuries were not only seeking to increase the volume and weight of the preparations they supplied to consumers, but also to end the need for patients to measure out the doses themselves and dilute the medicines to the required concentration. As a result, "vodkas" began to predominate over tinctures in pharmaceutical practice. A whole series of purely "Russian" factors conduced to this. The Russians not only lacked delicate, precise weighing instruments, but in general were unused to weighing small items in a precise manner. Russians also tended to be sceptical of anything that was small or scanty; they believed in the healing properties of big doses of medicine. Hence it was dangerous to entrust the preparation and measuring out of medicines to the patient, especially considering the extent of illiteracy in the sixteenth and seventeenth centuries, even among the boyars and gentry.

Although the term "vodka" was closely linked with medicinal

preparations, the procedure for obtaining *vodki*, that is, the dilution of redistilled or other spirits, could not fail to emerge as the general technological method of Russian distilling. Moreover, it was inevitable that this would be recognized quite early as a feature of the production of grain spirit in Russia. To perceive this it was only necessary to compare "Russian grain wine" with foreign liquors. Yet, as we shall see, this possibility did not arise before the seventeenth century and the Polish-Swedish intervention.

The Russian custom of diluting alcoholic drinks with water is so ancient, traceable from the Orthodox Church practice of diluting grape wine with water, through Byzantine tradition back to classical Greece, and is so obviously more necessary for grain spirit, with its much higher alcohol content, that it is safe to assume that the habit of putting water in vodka was established from the very beginnings of Russian alcohol distilling.

The term "vodka", nevertheless, was very much part of medical terminology, and for a long time did not pass into everyday use – let alone official language – with the meaning of an alcoholic drink. The reason for this was simply the strength of the medieval tradition (or, more exactly, rigid scholastic and psychological convention) which dictated that alcoholic drinks should be referred to as *vino*, qualified only on the basis of each drink's most obvious external characteristic – grape wine, grain wine, "service" wine or "burning" wine – and without any reference to its true nature or process of manufacture. To apothecaries, however, these technical characteristics were important, despite their being of no significance to patients – at least until the latter acquired the possibility of preparing the drugs in question and hence of becoming familiar with their technology. This possibility arose, however, only in the seventeenth century, when the state monopoly on the production of vodka was breached for a time, and the distilling of grain spirit was released from the strict government control of the sixteenth century.

As we have seen, in the sixteenth century vodka was effectively present both in medicines and among alcoholic liquors. Consequently, it went under different names: *vino* as a drink, and *vodka* as a medicine, though in their alcoholic content and method of their preparation there was little difference. This fact, naturally, could not remain hidden from those who produced vodka. In their everyday language these people would not have observed the formal distinctions, and would have used the word "vodka" to refer to the beverage as well as to the medicines prepared on the basis of vodka. Such loose terminology could, of course, only have been used as slang, in a particular social setting.

Written documents are encountered from the middle of the seventeenth century in which the word "vodka" is used to denote a drink. Thus in a complaint of Archimandrite Bartholomew from 1666 we read: "The old man Yefrem . . . now lives in a cell, and children secretly bring him drink, wine and *votka*." Here the term is used to denote a drink because of the need to distinguish between grape wine and "grain wine" – that is, vodka – within one sentence.

In this instance, the author of the complaint could not avoid using the slang word without distorting the content of his message. In all other cases when circumstances did not compel such usages, seventeenth-century writers continued steadfastly to use the term *vino* with the meaning of vodka. Thus in the seventeenth century the modern meaning was known in everyday language but was used only in particular circumstances.

It is important to note that as early as the mid-1630s many of the foreign travellers who visited Russia recorded the high quality and individual character of Russian vodka, although they described it using the terms employed for grain spirit in German or Swedish – *Branntwein* or *brännvin*. However, Johann de Rodes, the long-time Swedish resident at the court of Tsar Aleksey Mikhailovich, in his reports to Stockholm referred to Russian vodka by another term, *hwass*. Several historians have interpreted this as the result of a misunderstanding, supposing that Rodes had in mind *kvas*, or called vodka *kvas* through ignorance. However, the repeated use of this term in similar contexts means that it is entirely possible to identify it exclusively with vodka. Moreover, it is strange to suggest that a man so superbly well informed on the Russian conditions of the time might have made a mistake. From the point of view of Rodes the word *hwass* would have suggested the German *Wasser*, water (*voda*) rather than *kvas*, since for him the decisive consideration in seeking out resemblances could not have been Russian phonetics.

If we turn not to phonetics but to the meaning of the word in question, it turns out that in the Swedish language of the time *hwass* meant "sharp, biting, strong". However, in the lingua franca of the foreign residents of Moscow in the second half of the seventeenth century, a broken German, or Muscovite-German jargon, in which "Germans" from the Holy Roman Empire, the Hanseatic League, Livonia, Holstein and Holland expressed themselves, the word *hwass* or *hwassar* either had the same meaning as *Wasser* – that is, *voda* – or was as close to it as the Russian diminutive *vodka* was to *voda*.

If we consider this evidence, then it is clear that Rodes did not use this name for vodka by accident. He sought out a word which in Swedish meant strong, and which in the jargon of the foreigners in Moscow meant water. "Strong Moscow water" was a very successful translation of the word (and more importantly, of the concept of vodka), when that word had not yet entered any foreign language. The need to distinguish vodka from the German, Swedish and Ukrainian or Cherkassk varieties of "burning wine" was already apparent, because of the substantial differences in quality between these drinks and Russian vodka. These differences were connected with the use of particular materials, unknown in the West, for the purposes of purification. (This subject will be explored in more detail in Chapter 4.)

In the second half of the seventeenth century, therefore, both Russians and foreigners began to use terms for Russian vodka which distinguished it from similar drinks produced in neighbouring countries – Germany, Sweden, Poland and the Ukraine. It follows that vodka already possessed individual qualities which distinguished it from analogous grain spirits.

The word "vodka" began for the first time to be used in the official language in the eighteenth century, but as a secondary synonym rather than as a term of equal standing. Where the word appears without any explanations in official acts, handbooks and dictionaries, the note always follows: "See *vino*."

The only vodkas which were described without any reservations as "vodka" in the eighteenth century were those which were given an additional taste, aroma or colour; that is, those that underwent secondary distillation with the addition of herbs, berries, fruits, and sometimes even wood. Since multiple distillation was always indispensable for achieving the desired product, vodkas of this type were often described for the sake of brevity as *dvoennye vodki* (double vodkas), or took their name from the additive – aniseed, caraway, orange blossom, pepper and the like. All these vodkas were direct descendants of the medicinal vodkas of the sixteenth and seventeenth centuries, and made in a closely similar way. The sole difference was that they were infused not with medicinal substances, but with additives intended to impart a particular taste. This, however, was only a difference of form.

At the same time the medicinal herbal infusions were transformed during the "enlightened" eighteenth century into tinctures on the Western model. Where they were used as beverages, they acquired the name of *erofeichi* (from 1768). *Erofeichi* were not diluted with water, and therefore were not included in the category of vodkas. Their strength was of the order of 70 to 73 per cent alcohol.

At the end of the eighteenth century the word "vodka" was applied to three kinds of drinks. To distinguish between them the prominent Russian physician and natural scientist N.M. Maksimovich-Ambodik suggested in 1783 that the following terms be adopted: *vodki peregnannye* or *dvoennye* (distilled vodkas), *vodki nastoennye* (infused vodkas), and *vodki sladkie* (sweet vodkas).

These terms did not enter into the everyday language, but were occasionally used in the medical terminology of the time. In the nineteenth century plain distilled vodkas continued to be called *vino*, while vodkas sweetened with sugar or with the syrups of fruit or berries received the name *ratafii*, which came into general use in the second half of the eighteenth century and the first half of the nineteenth century, and was maintained throughout the nineteenth century. Aromatic vodkas, an invention of the eighteenth century, came to be referred to simply by the term "vodka", without the qualifier "infused". By the beginning of the nineteenth century "vodka" had thus come exclusively to denote aromatic vodkas prepared by the eighteenth-century method. So not only in the eighteenth, but also in the nineteenth century the only vodka that continued to be known as vino was the colourless, plain variety.

In the eighteenth century we very often encounter the term *vodka vpolu prostogo vina* or *vodka vpolu ot prostogo vina* (vodka half of "simple wine"). This term was used mostly to denote an intermediate product destined for use in the subsequent preparation either of a medicine or a beverage. It signified the addition of water in the volume of half the spirit obtained from the distillation of *raka*; that is, one-third water, a favourite ratio in vodka production. This product, whose strength was approximately 18 per cent alcohol, was not in most cases used as a drink. It generally served as a base for further distilling.

Thus the word "vodka" from the second half of the eighteenth century was used as a semiofficial term for plain vodka (*vino*), and as an official term for aromatic and coloured multiple-distilled vodkas. Meanwhile medicinal (herbal) vodkas lost their name and were transferred into the category of tinctures or *erofeichi*. In the second half of the eighteenth century the word "vodka" in everyday language also acquired overtones of slang, at times abusive and contemptuous, which can be linked to the spread among the common people of the worst kind of vodka, "vodka half of simple wine", which differed sharply both in quality and appearance from the high-quality aromatic vodkas produced for the gentry. The latter were known unofficially, in the colloquial language of the gentry and in the literature of the time, not as vodkas but as "infusions" (*nastoyki*). These terminological differences must always be kept in mind when we examine the history of vodka during the eighteenth century. Outside of the country aromatic vodkas were known from the end of the eighteenth century as "Russian vodkas", a term which was often used within Russia itself.

The nineteenth century witnessed the final stage in the conquest by the term "vodka" of what we now consider its primary meaning; that is, the term spread to include "grain wine", and not only the "Russian vodkas" which had arisen from the vodkas of the apothecaries.

It is difficult to establish how the term "vodka" acquired wide currency. The word "vodka" is in fact missing from the standard, respectable literary language of the first half of the nineteenth century. As evidence of this we can take the children's book on Russia by V. Buryanov, in which the author still calls vodka "burning wine".[10] Alexander Pushkin uses the word "vodka" quite often in his prose, but only with the eighteenth-century sense. In

Yevgeny Onegin Pushkin uses the term *russkaia vodka* to refer to anise vodka.

To judge from the dictionary of V. Dal, the word "vodka" appeared in the standard Russian language only in the 1860s, if we consider that the dictionary was compiled in the years between 1859 and 1872. Dal was not at all inclined to treat the word as a main headword. He treated it as a synonym for *vino* in the sense of spirits, and as a diminutive of *voda*. By this time, however, the primary meaning of "vodka" had in reality become the modern one.

When the word was first used in its present sense can be established only very approximately. At the beginning of the nineteenth century it is encountered in medical and culinary dictionaries as a fully valid term, though without a precise or correct definition. For example, P. Yengalychev in his *Domashnii lechebnik* (Home Medicine), published in 1825, notes: "1. Vodka or wine spirit. 2. Vodka or vodka infusions."[11] What is implied by the word in both cases is not the present meaning of this term. In the first case "vodka" is totally identified with alcoholic spirit, which is incorrect because a specific characteristic of vodka at this time was that it was a solution of alcohol in water. A linguist or philologist might not perhaps have paid any attention to this fact, but the compiler of a medical dictionary should not have failed to note it. If this point did not have any significance for Yengalychev, this indicates that he was not familiar with the concept of vodka in any detail, and suggests that the term had not spread to colourless vodkas. In the second case the term "vodka" was applied to "Russian vodkas" in the eighteenth-century sense. It follows that in 1825 substantial changes in the use of the term had still not occurred since the eighteenth century.

In 1838, however, in the largest and most detailed reference work of the first half of the nineteenth century, Plyushar's *Encyclopaedic Lexicon*, "Vodka" is included as a headword. The term is not explained, but is accompanied with a reference to an article on "Spirituous liquors". Since Plyushar's encyclopaedia was published only up to the letter D (V is the third letter of the cyrillic alphabet), we cannot establish how the authors of the work understood this word in 1838. It is clear, however, that by this time it was already a fully acknowledged term; the fact that it was included in the headword list of the encyclopedia means that it was no longer viewed as slang. (It should be explained that the listing of Plyushar's dictionary was compiled by N. Grech, who edited out anything which might have shocked Nicholas I, who in 1836 had agreed to serve as the patron of the *Encyclopaedic Lexicon*.) In the article on "Distilling" included in the same volume the term "vodka" is absent, but "grain wine" is characterized as alcoholic spirit diluted with water; that is, it corresponds to the concept of vodka, though this term is not employed.

Finally, in the *Economic Lexicon* of E. and A. Avdeev, published in 1848, "Vodka" is again used as a headword, but is understood only to refer to sweetened or aromatic types (there are references to mint, caraway, angelica and so on). The recipes cited for the preparation of these drinks do not in all cases envisage the dilution of grain spirit with water; both vodkas and tinctures are included under the entry for "Vodka". This might, of course, be ascribed to the incompetence of the authors; but, since we are dealing here with the most widely known and authoritative reference work of the mid-nineteenth century, we can conclude that by this time the term "vodka" had come to be used more broadly, that it had entered fully into the language, and that it had ceased to repel or shock educated people with its aura of slang. At the same time, however, the clear terminological significance of the concept was becoming blurred. The associations that the word had carried in the eighteenth century had gradually been lost, at least in part.

At any rate, both Plyushar and the Avdeevs confirm indirectly that the former view, that vodka of its essence had to include water and that a spirituous liquor without at least a third of water was not vodka, had completely disappeared. In the *Economic Lexicon* vodkas are taken to include *erofeichi*, which were characterized precisely by the fact that they were not diluted with water.

Thus what occurred during the first half of the nineteenth century was merely an increase in the currency of the word "vodka" itself, in the sense that it lost its overtones of slang. The concept of vodka not only failed to spread, and failed approach its modern meaning, but also became distorted in relation to the significance it had possessed in the eighteenth century. The concept did not become transferred to vodka proper; that is, to grain spirit diluted with water.

It should be noted that throughout the nineteenth century the expression *dvoennaia vodka* (double vodka), which had first appeared in the 1730s, remained in use. For some time it was employed with its correct meaning. Evidently in consequence of the fact that "double" spirits – that is, strong spirits – were always diluted with water, the expression "double vodka" proved extremely durable. The word *dvoennaia* was always accompanied by the word *vodka*. But by the 1860s the meaning of this phrase had also been lost, since from a number of literary references we know that *dvoennaia vodka* implied only secondary-distilled spirit, not diluted with water. So at the beginning of the second half of the nineteenth century the word "vodka" was becoming implanted in the language while simultaneously losing its definite, specific meaning.

As a result, by the 1870s the word "vodka" had begun slowly to oust *vino* (in its sense of grain spirit) from the spoken language. At the same time, two factors were hindering this process. The first of these was the language of government administration, according to which all the trading and catering establishments of Russia dealt in a product named *vino*. The second factor to play an important role here was conservative social consciousness – the view that the word "vodka", when used to refer to grain spirit, was only an expression of the lower strata of society. As a result of this prejudice the middle classes, which were growing in size and importance after the abolition of serfdom in 1861, avoided where possible using the word "vodka" in their affectedly proper speech. Particular evidence of this is supplied by the fact that of 400 proverbs devoted to vodka, only three call it by this name, while all the rest either refer to it as *vino* or employ various euphemisms. The three proverbs that do use the word "vodka" are all new ones, dating from the period after the 1861 reform.[12]

The main importance of these sayings is that they establish clearly the connection between the name "vodka" and the drink which people until this time had been accustomed to call "grain wine"; as a consequence, the proverbs return the name to its real owner. It is also important to note that thanks to the fact that these proverbs can be dated fairly precisely to the end of the 1860s and the beginning of the 1870s, they provide an additional basis for identifying this as the period when the name "vodka" took over from "grain wine" as the standard Russian usage. We can therefore conclude that vodka proper began to be called "vodka" around this time.

However, the use of the term was still far from universal. The word was strongly implanted only in Moscow and its environs, and to some extent in St Petersburg; while in various other regions it was almost never heard until the end of the nineteenth century. Thus, to the east of Moscow in the provinces of Vladimir, Nizhny Novgorod, and to some degree in Yaroslavl, Kostroma and Ivanovo, the implantation of the word "vodka" with the meaning of an alcoholic drink was hindered by the local habit of using the word with the sense of water ("Go to the stream and fetch some *vodka*!"). In Arkhangelsk and Vologda provinces and in northern Karelia, and also to some extent in the provinces of Novgorod and even Tver, the word continued even in the second half of the nineteenth century to be used with its old Novgorod meaning of bustle or useless traipsing to and fro, a usage derived from the verb *vodit'* (to lead). As a result, through the 1890s and until the complete introduction of the monopoly in 1902, two names for vodka continued to exist, *vodka* and *vino*. New euphemisms also emerged – *belenkoe* and *beloe* (white, where the use of the neuter gender still implied *vino*) and m*onopolka* and *popovka* (where the use of the feminine implied "vodka") – while in official language the term *vino* predominated right up to 1906.

Meanwhile from the 1870s to the 1890s, especially in the years directly preceding the introduction of the state monopoly on vodka, there were attempts to create and bring into use completely new and artificial names for vodka, with the false aim of ennobling the product. Examples of such names included "people's wine", "table wine" and the totally confused "wine vodka". None of these took root; they all proved alien both as elements of the Russian language and as expressions of the concept of vodka.

D.I. Mendeleyev, who took an active part in establishing a modern, scientific technology for the production of vodka, decisively rejected all these artificial names and insisted on the introduction of a single official designation, vodka, which expressed the nature of the drink most exactly and which at the same time was the most authentically national Russian name.

During the 1880s and 1890s it was the practice to describe as vodka spirituous liquors whose alcoholic content varied from 40 to 65 per cent, while liquors containing from 80 to 96 per cent began to be called spirits.[13] Then in 1902 a law was enacted under which vodka with the ideal strength, that is, 40 per cent alcohol, could be termed true vodka – Moscow vodka.

From the mid-1860s to the reintroduction of the monopoly on vodka between 1894 and 1896 vodka was usually formulated on a very simple basis – a mixture of 50 per cent of alcohol with 50 per cent of water. This ratio succeeded the former proportions of 1:2, and gave a drink containing 41 to 42 per cent of alcohol by weight. The fact that the percentage of alcohol finished up at this figure was the result of a curious phenomenon: when alcohol is mixed with water the total volume of liquid is reduced. This means that if we take one litre of pure water and mix it with one litre of 96 per cent spirit, we finish up not with two litres of liquid but with considerably less. The stronger the spirit, the greater this diminution.

These phenomena were observed by Mendeleyev, who began to study their connection with changes in the quality of mixtures of water and spirit. It turned out that the physical, chemical and physiological qualities of these mixtures were also extremely diverse, encouraging Mendeleyev to search for the ideal proportions. While the earlier practice had been to mix various volumes, Mendeleyev mixed samples on the basis of the weight of water and spirit, which was more difficult but gave more exact results. It turned out that the ideal alcohol content for vodka was 40 per cent by weight, which had never been obtained in a precise way through mixing by volume. From this time (1894–6) Russian vodka – or, more precisely, Moscow vodka – came to be considered as a product consisting of grain spirit, triple-distilled and then diluted with water to a concentration of 40 per cent by weight. Mendeleyev's formulation was adopted in 1894 by the Russian government as the standard for the country's national vodka.

Meanwhile, in all the countries of Western Europe, spirit obtained from diverse raw materials (sugar beet, potatoes, plums, barley, wheat, palm juice, sugar cane and others) was usually diluted in the proportion of 1:1 by volume. "Vodkas" were obtained with alcohol contents such as 42.48 per cent, 42.1 per cent, or 30.7 per cent; that is, which either exceeded or did not attain the "golden mean" of 40 per cent. This immediately set the new Russian vodka apart from all of the "vodka-like" alcoholic liquors in other countries, which differed sharply from Russian vodka in principle, and not simply in quality.

Vodka in its modern form owes much to Mendeleyev's discovery. Yet that does not diminish in any way the importance of the previous history of the development of vodka, for its perfection at the end of the nineteenth century was the logical result of its previous evolution, the result of the vast empirical labour which allowed Mendeleyev to define the ideal form of Russia's national drink.

4

Vodka Production and Its Control

The Evolution of the Technology of Vodka Production

In a work in French on the history of winemaking one of the most famous of contemporary Italian oenologists, Dr Oberto Spinola, uses in place of the usual *eau de vie* – equivalent to *aqua vitae* – a more precise term, *eau de vie de vin* (water of life made from wine).[1] In this way he lays stress on the fact that the spirit obtained at the beginning of the fourteenth century by Arnold de Villeneuve was distilled from natural grape wine. It was extremely pure and delicate, and light in flavour. It was, in fact, not mere alcohol in the current chemical sense of the term; but a finished product, a fine brandy similar to cognac, containing besides ordinary ethyl alcohol a mass of other compounds which had passed with the spirit through the still.

It is easy to see why the raw material base of this spirit not only helped determine the design and construction of the distilling apparatus, but also influenced the subsequent processing of the product that was obtained. The pleasant taste and aroma of this product called for almost no additional processing or refining, apart from maturing in oak barrels, which softened the flavour (in the process allowing some of the spirit to evaporate), removed some of the more harsh-tasting and poisonous congeners (substances such as heavy alcohols which had gone through the still with the ethyl alcohol), and imparted a beautiful golden colour.

In the early Middle Ages *sikera* – in the sense of a raisin, date or fig vodka – had been prepared in the Middle East, mainly in Palestine and Asia Minor, by Jews, Byzantine Greeks, and Arabs. This was a considerably cruder beverage than French brandy, since it was made from a liquid fermented from hard sun-dried fruits, not from their juices. Nevertheless it had some of the flavour of natural fruits, and had almost no need of complex refining or additional processing; multiple distilling gave an exceptionally pure product with a delicate aroma. The fruit base was largely free of starch; in the case of grape wine the sugar that was present was mostly in the simple forms of dextrose (glucose) and fructose, on which the yeast could work directly, so that little residue was left in the liquid. The result was to make possible a high degree of purity in these, humanity's first spirits. In the preparation of these drinks the question of using special methods of purification did not even arise. It is therefore not surprising that the fruit and grape brandies received the name "water of life". Stimulating, enlivening effects were observed to flow from their use; only gross overindulgence would produce painful consequences.

The raw material for grain spirit was a totally different substance, wet dough, which demanded different equipment. The transformation of starch into sugars created other chemical compounds as well, and the liquor that resulted was only an intermediate product, highly contaminated with congeners and with a far from pleasant taste and aroma. One of the principal contaminants was – and in badly made vodka, still is – fusel oil, a smelly and noxious mixture consisting largely of butyl and iso-amyl alcohol. From earliest times this problem inspired attempts at further processing and refining.

Thus from the very first a clear boundary arose between grape spirit, together with its close relatives the other fruit spirits, and spirit obtained from the starch in cereals – that is, vodka. Consequently the development of their distillation – and later that of milk "vodkas", the so-called *arki* – could not occur through direct borrowing, and had to be conducted largely from first principles in lands whose people used cereals, or milk, as the raw material for alcoholic beverages.

In Southern Europe, Asia Minor, the Mediterranean region and

Transcaucasia the production of cognac and of local fruit vodkas arose and could arise only out of winemaking, from viticulture and the tending of orchards. In Russia and in Northern and Eastern Europe, on the other hand, the rise of vodka was rooted in cereal culture. Vodka production arose naturally out of brewing, which from time immemorial had used grain as its raw material, and whose practitioners considered alcoholic drinks largely in terms of grain products.

In exactly the same way *arka* or *kumyshka* (milk vodka) arose among herding peoples of Asia and Eastern Europe who possessed a surplus of milk, and who used it to prepare a variety of milk products, including *kumys*, a form of yoghurt made alcoholic by naturally present wild yeasts. It was through introducing new technology in the course of attempts to create a new type of cheese that they obtained *arka* as a by-product.

Thus the discovery of brandy, vodka and *arka* was in each case necessarily an independent development, whose technical side was closely connected with the raw materials available in each region, and with the experience in their processing which had been acquired during earlier stages of history.

For a whole variety of reasons, relating above all to manufacturing and trade secrets, this section will not be able to describe the technology of vodka production in detail. Nor will it concern itself with the modern technological process or give other details connected with the production of vodka at the end of the nineteenth century or in our own time. Nevertheless, it is essential to give a reasonably clear idea of the production methods that were used in various periods, and which can be regarded as aspects of the creation of vodka as a beverage. This will familiarize us with the drink on a scientific, cognitive level, and help us to understand that we are dealing with a complex historical, cultural and technological phenomenon.

The factors that were decisive in establishing the original nature of the drink, and continued to figure in the prolonged historical process through which vodka was developed, were raw materials, formulas, methods of purification, distillation techniques, and equipment. The roles which these played at different historical stages were far from uniform.

During the first stage, from the fifteenth to the seventeenth century, the weakest link was equipment. This weakness had a powerful effect on the choice of techniques. In the West the main factor which allowed progress to be made in the production of alcoholic spirit was the perfecting of the equipment and of the basic process of distilling. The Russians, on the other hand, always relied on the distinctive qualities of the raw materials and of their combination. The imperfections of their equipment forced them to use elaborate methods for refining the end product of distillation, and it was in this direction that technological development proceeded.

Methods of refining vodka acquired special importance in the eighteenth century, when the gentry oversaw a period of rapid development of distilling, and purifying processes were introduced regardless of cost. This was also an extremely fruitful period for research and experiment into the raw materials and formulas for vodka, with the introduction into the mash of various aromatic ingredients.

It was only in the second half of the nineteenth, and especially in the twentieth, century that attention came to be concentrated on the apparatus. Modifications were made to the way it functioned, and attention was paid to such purely technical questions as the temperature and speed at which the various productive processes went ahead. To improve the quality of the raw materials, methods which had not even been dreamt of in earlier times, such as the aeration of water, were introduced.

All this contributed to the creation of modern vodka; that is, of vodka which is not simply a means of getting drunk, but a complex national product embodying the historical and technological imagination of the Russian people.

It was for this reason, more than popular consciousness or Russian tastes, that it proved so difficult to wean the nation off vodka when in 1986 the official decision was taken to reduce production almost to the point of abolishing it. Vodka had emerged through a historical process, and it could cease to play its current role in the life of the individual and society only in accordance with the logic of historical development. It was stupid and presumptuous, at the very least, to suppose that anything so rooted in the national culture could be abolished by a mere decree.

Here then, in summary form, are the basic technological characteristics of vodka and of its production.

Raw Materials

Grain: for centuries, until the 1870s, rye served as the basic raw material for the production of Russian vodka. Since then, and especially since the 1930s, wheat has come to play a much greater role in the production of the popular brands of vodka, and in times of war or economic collapse potatoes and other crops have been used. However, the finest varieties of vodka continue to this day to be produced from rye grain and rye bran. While rye is the essence of true Russian vodka, the process also uses other cereals – oats, wheat, barley and buckwheat – in proportions which vary but are always small.

This grain base, and especially the use of rye, ensures that Russian vodka is greatly superior to potato vodka; something which was already noted by Friedrich Engels. Russian rye vodka does not induce such after-effects as a heavy hangover, and does not create aggressive moods in the drinker in the same way as potato or, worse still, sugar-beet vodka.

Water: the second most important ingredient of vodka is water, or more precisely, the soft water of Russian rivers. To be suitable for making vodka, water needs to have a hardness of no more than 4 milligrams equivalent per litre of dissolved minerals. Until the 1920s the water of the Moscow river (2 milligrams) and of the Neva river (4 milligrams) met these requirements. The best-quality water was and remains that of the Mytishchi springs, from which water was already brought to Moscow in the eighteenth century, over a distance of more than twenty kilometres. At present the water for Moscow vodka is brought from the Vazuza, a tributary of the Moscow river, in its upper reaches to the west of Moscow in the Gzhatsk (Gagarin) district of Smolensk province. The Vazuza flows through a thickly forested region and has pure, soft water (from 2 to 3 milligrams equivalent).

Before it is blended with the grain spirit, the water undergoes a sequence of processes to purify it further. It is allowed to stand, and is filtered through two grades of sand. It also undergoes a special additional aeration, in which the water is saturated with pure oxygen. In no circumstances is the water subjected to boiling or distillation, as is normally done by the producers of pseudo-vodkas in other countries (the USA, Finland, Germany and elsewhere). This is an important distinguishing feature of Russian vodka, and one of the sources of its superiority. Russian vodka possesses a special softness and smoothness, since the water in it is not soulless but living; it has no noticeable odour or aftertaste, but equally it lacks the flat, stale taste of distilled water. The degree of purity of this Russian water is such that it possesses a crystal transparency, measurably greater than that of distilled water, which has been deprived by the distillation process of its natural sparkle and brilliance.

Malt: this is an important ingredient used in the preparation of the mash. Russian malt has always been made from rye. Even at the beginning of the twentieth century, when wheat began to emerge as the basic grain used in the fermentation process, and in the period from the 1930s to the 1950s, when for economic reasons production in the Soviet Union of simple, cheap potato vodkas was increased, rye malt was always employed. The way in which the malt is prepared, especially the conditions under which the grain is allowed to sprout, is a factor of significant and even decisive importance for the quality of traditional Russian vodka. As early as the eighteenth century Academician Tobias Lovits and the landowner and distiller V. Prokopovich studied the methods used in preparing rye malt for distilling, and made precise recommendations.[2]

Yeast: Russian distilling originally employed the same rye leaven used for the baking of black rye bread. During the eighteenth century there was a general shift to the use of beer yeasts, which were more active and speeded up the fermentation of the mash.

From the late nineteenth century special pure strains of yeast were prepared in distilling works. These were intended specifically for vodka production, and were added to the wort in the fermentation vats. The correct maturation of the mash, and hence the overall quality of the final vodka, depended heavily on the quality of the yeast.

Formulas

The formula for the mash, including the proportions of grain, water, malt, yeast and additional aromatic substances, has always been an object of research and experimentation by Russian distillers. The additives which have been used include forest herbs, the young buds of various Russian trees (birches, willows and pussy-willows), leaves (cherry and blackcurrant) and foreign spices (amaranthus, cloves, mace and others). The range of substances used expanded greatly in the eighteenth and the first half of the nineteenth century.

The most characteristic feature of the Russian formula is, without doubt, the addition to the basic rye grain of small but vital quantities of barley, buckwheat, oat flakes, wheat bran and cracked wheat – that is, the remnants which used to accumulate in mills and in the farm establishments of large landowners as a result of the processing of various grains. At first these ingredients were not added at all methodically; but it was noted that despite making up only one or two per cent of the total weight of the grain component of the mash, they were capable of giving the vodka an elusive but perceptible individual character, of imparting to every batch its particular "face", yet without altering its traditional qualities. At the end of the eighteenth century academicians working in the fields of chemistry and botany became interested in these empirical observations of Russian distillers. The scientists conducted their own experiments, and published a series of works recommending the use in vodka production of various small additions to the rye which formed the basic raw material.[3]

In Chapter 3 we touched on the strengths of alcoholic spirits ("simple", "double" or "triple"), and on the numerous mixtures of which they have formed a part. This shows clearly that the road to the modern proportions of spirit and water in vodka, measured by weight as established by D.I. Mendeleyev, was a long one. It passed through a number of stages, during which various experiments were tried, including unsuccessful ones ("wine with a wave"). The experiments had as their starting-point the traditional Greek and Byzantine practice of adding two parts of water to one of wine, and culminated in an ideal mixture, expressed in precise mathematical terms: 40 per cent by weight of pure spirit and the rest water. But that water must be the pure, soft water of small Russian forest streams, which cannot be reproduced anywhere in the world.

Methods of Purification

From the very beginning of vodka production, methods of purifying the drink have had an important place. The development of these methods has no analogy in the production of alcoholic spirits in Western Europe. The age-old preference of the Russian drinker for aromatic drinks of the mead and beer type forced the original fifteenth-century distillers, whose primitive distilling process and equipment yielded a foul-tasting spirit with a repulsive odour, to try every possible method of ridding their product of these qualities. The distillers therefore sought in the first instance to find effective ways of purifying grain spirit of these contaminants – substances which we now know by generic names such as oils, esters and aldehydes. Perfecting the distillation process was impossible, since the distillers did not possess properly sealed copper or glass stills, and had no hope of obtaining equipment made of such refined materials. Their hopes thus resided in the methods of improving quality which had been tested out in mead brewing and maturing, and which were now applied to vodka. There were several techniques of very different kinds.

Mechanical methods   These were the first to be used. The newly produced spirit might be rapidly cooled. *Raka* was put out in the frost immediately after distillation. After standing for a while, the spirit might then be decanted into another container. Any impurities which had congealed in the cold or sunk to the bottom would be left behind. (These two operations were transferred to distilling from the preparation of matured mead. They are also often practised in winemaking.)

Filtration of the raw spirit, of the water-spirit mixtures and of the vodka, were all used; and filtration remains vital today. In Russian distilling the filtration processes were carried out with extreme care over a prolonged period. Knowledge in this area was accumulated gradually and handed down as a trade secret from one generation of distillers to the next. By the beginning of the nineteenth century lengthy empirical observation had brought the process to a high level of refinement, but it nevertheless continued to be improved throughout the nineteenth and twentieth centuries. Various materials were used as filters: felt of various types; woollen, linen or cotton cloth or cotton wool; paper of various thicknesses and densities; river, sea and quarried sand; crushed stone; broken pottery; and charcoal – in the twentieth century, activated charcoal, which is treated to increase its absorbing power.

In the history of Russian distilling, filtration through charcoal occupies a special place. Russian distillers discovered empirically one of the basic rules that ensures the special qualities of Russian vodka. This rule is that raw spirit or any other kind of pure spirit must not be filtered directly through charcoal, which cannot remove oily contaminants from highly concentrated spirit. Instead, it is essential to dilute the spirit with water to an alcohol content of at most 55 per cent, ideally 40 per cent. From the end of the eighteenth century special attention was devoted to raising the absorbing properties of charcoal through the preparation of the wood designated for this use. To improve the initial quality of the wood, various steps could be taken. Among these, the bark might be removed from the wood before the latter was burnt; knots might be cut out; heartwood, especially that which was darker than the rest of the wood, and also the outermost layer, might be discarded; and trees older than forty or fifty years might be excluded completely.

Finally, it was established by empirical methods that charcoal from different types of trees has different absorbent capacities. It is far from being a matter of indifference which type is used to filter the finest varieties of vodka. In order of absorbency from highest to lowest, these woods are beech, lime, oak, alder, birch, pine, fir, aspen and poplar.

The first four types are expensive, and were used mainly during the eighteenth century in the home distilling of the gentry, and to some extent in the nineteenth century for the production of the finest varieties of vodka. Apart from this their use was limited to particular geographical regions. Alder charcoal was used in private distilling until the year 1861. Limewood charcoal was used as late as the Soviet period, until 1940. But from a very early stage, in the fifteenth century, most of the charcoal used in Russian distilling has been made from birchwood. This is the cheapest and most widely available; throughout the nineteenth century it was produced on a huge scale as a domestic requisite, for it was also used as a fuel for samovars. Birchwood charcoal is not far behind the first four in absorbency.

The effectiveness of the simple birch charcoal filters used in Russian distilling in the nineteenth century, before the discovery of activated charcoal, is shown by an experiment performed in the 1880s. The experimenters began with grain spirit, in which standard chemical analysis had been unable to detect even a trace of aldehydes. After the spirit had been diluted with water to 45 per cent alcoholic strength and filtered through granulated birch charcoal in four columns, as much as 0.11 per cent of aldehydes was retained in the filter. The spirit was thus refined to virtual purity, when it was impossible to detect any trace of aldehydes even by an ultra-sensitive test with rosaniline, which changes colour when added to a solution containing even one thousandth of one per cent of aldehydes.[4]

Chemical methods   These began to be used, in combination with mechanical methods, at a relatively early stage in the production of vodka, in the seventeenth and more widely in the eighteenth century. They were highly effective, especially in ridding vodka of unwanted odours.

Coagulants might be used in the distilling process. Both raw spirit (*raka*) and other distillates were treated with natural biochemically active coagulants, which solidified around the contaminants so that these would be left behind in subsequent distillation. Examples of such coagulants were milk, whole eggs and egg whites. Sometimes freshly baked black bread was added; this also traps contaminants. It was used to purify double-distilled grain spirit, which had been treated with milk at the first stage.

Of course, these natural methods of purifying vodka made it much more costly, especially since their solidification made it impossible to distil more than 45 per cent of the mash. But in the mixed enterprise of the landed estates the leftovers from the mash, including the expensive eggs, bread and milk, went to feed stock, and were not completely wasted. Furthermore, such refining gave a product of exceptional purity and delicacy.

Other purifying agents used by distillers included ashes, potash, burnt wormwood, and later soda, which were added to double or triple-distilled spirit. These mixtures were used to achieve the very highest degree of purification, in the production of quadruple-distilled spirit or absolute alcohol (that is, of almost 96 per cent purity, the highest strength that can be produced by distillation alone).

However, high-grade distillation could not be achieved everywhere, and distillation in itself still did not provide any guarantee that the product would be free of odorous and oily contaminants. Therefore, such means of purification as chilling and the use of isinglass were taken over from winemaking, and were often applied not to the intermediate but to the final product, after it had been diluted with water.

Chilling was cheap though wasteful, and produced an excellent result. Thanks to the heavy Russian frosts, and also to the practice of storing huge blocks of ice through the summer, chilling large batches of vodka was not difficult. The vodka was frozen in special small kegs with removable bottoms or with special plugs through which the spirit that had not been frozen was poured off. The water which had remained in the vodka was transformed into a block of ice; oily contaminants hardened on top of this or settled in a thin layer on the walls of the cask. They were thus easily separated and discarded.

The other operation, the use of isinglass, was rather expensive, but did not require a great deal of time and had a more precise effect. With this method it was possible to remove with great throughness odorous substances and other contaminants from grain spirit or already prepared vodka. Isinglass is a powerful adhesive: the impurities literally stuck to it. At the end of the process the isinglass and its load of contaminants were removed by filtration through cotton cloth.

Flavouring Originally used to conceal unpleasant flavours, but later simply to impart an interesting taste, Russian distilling also employed various means of flavouring vodka which had been inherited from the ancient traditions of mead-brewing. Initially, hops and herbs – known collectively as *zel'e* – were used to give the vodka added aroma and, it was then believed, strength. In the eighteenth century the juice of wild fruits (rowan berries, raspberries and strawberries) was added. Eventually these practices led to the development of a separate branch of the distilling industry, creating an array of "Russian aromatic vodkas" and liqueurs.

The Equipment and Technology of Distilling

If we take into account all the additional processes used to purify the raw materials, the intermediate products and the final product, the technology employed over the centuries to produce alcoholic spirit in Russia has been quite unlike that used in Western Europe. Especially during the earlier years, the equipment employed in Russia has also had its own distinct characteristics.

In Chapter 1 it was noted that no examples have been preserved of the earliest equipment used in the production of vodka. The main reasons for this are the simplicity of the equipment, its short useful life and its frequent replacement. The materials from which the apparatus was made – kiln-fired clay and wood – were not particularly durable. Nevertheless it is possible on the basis of indirect data, including written sources and linguistic evidence, to distinguish three stages in the development of the equipment used in Russian vodka production.

The first stage, in the fifteenth century, involved the "pot" method of obtaining alcoholic spirit, which has been described – as far as possible from the scanty information available – in Chapter 1. This method was really no more than a concentration of the less volatile elements in the original brew. There were no pipes to catch and condense the light, volatile ethyl alcohol (the normal constituent of alcoholic drinks), which must mostly have been lost, so that the intoxicating effect of the drink was from heavy, poisonous congeners which did not boil off so readily.

The second stage, from the early decades of the sixteenth century, saw the use of a primitive pipe method. After the first fermentation in wooden vats, large jugs or pots of fired clay were used. These did not come into contact with the fire, but were heated in Russian stoves. The ethyl alcohol vapour was led off through clay channels or through pipes fashioned from the outer bark of birch trees. This vapour condensed when it passed through these channels or pipes into a neighbouring unheated area, a porch or shed. Distilling was usually carried on in winter, when the harvest had been gathered and it was possible to work out how much of the grain could be used for the production of vodka. During these frosty months the population had ample time for the slow process.

The seventeenth century saw the introduction of directly-heated copper vats and of copper pipes for condensation. This equipment was imported from Poland, Livonia, Germany and Sweden, and was in use from the 1620s, from the period of the Swedish–Polish campaigns.

In the eighteenth century Western European distilling equipment was introduced, first at the Academy of Sciences in St Petersburg and then gradually throughout the whole country. Tradition demanded, however, that characteristically Russian methods of purifying vodka should still be used.

From the middle of the nineteenth century the transition took place to factory distilling, employing German and British technology. This included Henze's system for handling mash, and his heating apparatus; Delbrück's open vessels; the distilling equipment of Selje-Blumenthal and Pistorius; Coffey's still, which could separate light from heavy distillates; and other devices. The conditions that existed in Russia quickly exposed the shortcomings of such equipment – its large size and high cost, the large quantities of metal needed for its construction, its tendency to become clogged, and the impossibility of using it to distil spirit from a thick mash. Improvements were needed, and various innovations were made, so that by the end of the nineteenth century Russia had established its own forms of distilling apparatus. These were modernized and further improved in the period from 1924 to 1941, and also during the post-war decades.

The actual process of distillation was no different in principle from that employed elsewhere. Until the 1870s, however, Russian distilling was characterized by two main rules: that it should proceed as slowly as possible, and that not more than 45 per cent of the volume of the wort should be distilled off. These rules were applied not only to the first distillation, through which *raka* and "simple wine" were obtained, but also to subsequent distillations. These practices naturally led to significant losses of raw materials and intermediate products, and were possible only under the non-commodity, non-market conditions of the Russian serf-holding economy. The decisive consideration here was not the profitability of a particular type of production, but the quality of the product; the vodka-producing landowner took no heed of costs or losses as long as a high-quality product was obtained.

As an illustration we shall take just one figure. It required 1,200 litres of mash, containing 340 litres of grain and rye malt and 12 litres of beer yeast, to produce only 3.5 buckets (42 litres) of good "simple wine". Redistillation of these 42 litres, with the then obligatory addition of about a bucket of milk, gave a maximum of 15 litres of good, pure grain spirit. From this the landowner-distiller, adding the traditional third part of water, obtained a total of 20 litres of top-quality vodka. For a landowner who received grain free of charge from his peasants, who had firewood from his own forests and who obtained the labour of the distilling workers virtually for nothing, a yield of vodka amounting to less than 2 per cent of the volume of mash was nothing alarming, and was not considered a loss or a heavy expense.

The whole process of production was aimed at satisfying the whims of the landowner and of his guests, not at making profits through the sale of vodka or at turning vodka into a market commodity. The governments of Peter I, Elizabeth I and Catherine II, which granted and constantly extended the privileges of the gentry in the area of domestic distilling, freeing the gentlemen-distillers from all controls and taxes, nevertheless consistently stressed that all the vodka produced had to go directly for the personal, household and family use of the gentry, and that it must in no case become an object of trade. The gentry in turn gave their solemn promise to the monarchs that they would honour the status of vodka production as a privilege of their estate, and would not attempt to turn it into a vulgar source of enrichment. It was thus in the social context of serf-holding Russia that vodka attained its very highest quality and acquired an extremely diverse assortment of varieties, the differences between which were often barely perceptible, but were jealously maintained.

The vodka produced in the aristocratic households of the Russian magnates – the princes Sheremetev and Kurakin and the counts Rumyantsev and Razumovsky – was of such a high standard that it put even famous French cognacs in the shade. This is why Catherine II did not hesitate to bestow gifts of vodka on monarchs of the order of Frederick the Great and Gustav III of Sweden, not to speak of petty Italian and German rulers. She also sent this refined and exotic liquor to Voltaire, who was well schooled in French wines. Catherine did not fear in the least that she might become a victim of Voltaire's murderous sarcasm. To the doubts expressed by a courtier about whether it was appropriate to give vodka to a philosopher, and especially to one with as sharp and unrestrained a tongue as Voltaire, Catherine is said to have replied condescendingly: "After this he'll swallow his tongue, whether from surprise or delight, or out of envy for Russia." Of the intellectual luminaries of the time, it was not only Voltaire who received this valuable gift, but also Carl Linnaeus, Immanuel Kant, the Swiss scientist and poet Johann Kaspar Lavater, Johann Wolfgang von Goethe, and many others. It is well known that the great chemist Mikhail Lomonosov esteemed vodka somewhat too highly, but that did not detract from his status as a scientist; and if anyone knew all the secrets of purifying vodka, it was he. When Linnaeus encountered vodka he was so inspired by it that he wrote a whole treatise, *Vodka in the Hands of the Philosopher, the Physician and the Common Person.* This is a highly interesting work, in which the great scientist sets forward a broad and objective social, medical, economic and moral evaluation of this product.

In sum, the high quality of the "domestic" vodka produced in the households of the gentry won it international prestige as early as the eighteenth century. This vodka became the drink of the flower of society, a product with an outstanding reputation for its purity and for the medical benefits of its use. It is true that there were people in the eighteenth century who became drunkards on vodka. A well known example is the poet Ivan Barkov, the father of Russian pornographic verse. He began drinking while still a schoolboy, and drank nothing apart from cheap tavern vodka. In the eighteenth century this was not noted for its purity; in effect it was simply *raka*, not even the ordinary "simple" grain spirit. It is no wonder that Barkov, who at the age of sixteen was as physically developed as the twenty-seven-year-old Lomonosov, quickly sank into alcoholism and died at the age of thirty-five. The fact is that vodka in the hands of the philosopher, the physician and the common person acts in completely different fashion, the more so since the vodkas which correspond to these diverse social categories are completely different.

Where the quality of vodka was concerned, the development of capitalism in Russia was disastrous for the common people. The quest for wealth brought cheap brands of Ukrainian potato and beet vodka, mainly from Kiev and Poltava provinces, on to the Russian market. These vodkas were sold only on draught and by the bucket, which led to the most unrestrained drunkenness. The production of pure, high-quality vodka was unprofitable for the capitalists, who were completely set on the production of vodka for sale. Moreover, in the second half of the nineteenth century Russian rye vodka began systematically to be exported to Germany, with the result that the mass market within Russia came to be dominated still more heavily by the potato vodkas. This forced even the Tsarist government to recognize that the market and market relations in Russia were unable to regulate the quality of vodka, and to decide that it was essential to reintroduce centralized production and trade in the drink, with strict government controls to thwart possible abuses. This was how the introduction in the years from 1894 to 1902 of the vodka monopoly was justified. The fact that this policy of strict state controls was continued and applied consistently after the Revolution of October 1917 has saved vodka, as a product prepared to a high standard, from deteriorating.

Throughout the years since the Revolution the Soviet vodka factories made use of the technological studies prepared by the commission which oversaw the introduction of the vodka monopoly between 1894 and 1902. Those who conducted these studies included scientists and engineers such as D.I. Mendeleyev and N. Tavildarov. The Soviet distilling industry has put out a product of the same high quality as the best state enterprises of pre-Revolutionary Russia. Meanwhile the vodka firms founded in

Europe and the USA by the White generals, bankers and industrialists who fled Russia have possessed neither Mendeleyev's
technical studies, nor the original Russian and Soviet equipment
designed specifically for producing vodka. These firms have based
their operations on typical Western European and American distilling apparatus. Consequently they bring out a cleanly distilled,
elegantly packaged product, but one which lacks the characteristic traits of Russian vodka. In other words, these are not vodkas
but pseudo-vodkas, since in terms of their raw materials, their
technology, and even such a cheap ingredient as water, they dif-
fer sharply from Russian vodka. Even the exquisite Finnish vodka
Finlandia, which uses rye grain and rye malt exclusively, has a
taste which is distinctly different from Moscow vodka. Unlike
other foreign vodkas Finlandia is extremely natural, and there is
no doubt about the use of rye in its preparation, since the
Finnish makers are utterly scrupulous. However, Finlandia still
cannot stand comparison with Moscow vodka. This is because
Finnish vodka uses Vasa rye, the grains of which are heavier and
cleaner than those of Russian rye, and which do not possess the
characteristic rye taste of the Russian cereal.

It is notable that Vasa rye completely deteriorates and shrinks
after three or four generations of seed, while our normal Russian
rye, despite all the vagaries of the weather and other afflictions,
maintains its standard unchanged not only for decades but for
centuries. Hence, for purely biological and geographical reasons
it is impossible to make Russian vodka anywhere outside the borders of Russia. One might reproduce the equipment and the technical process; but it is impossible to create artificially, somewhere
in Illinois or Cheshire, the soft water of the Russian forest river
Vazuza or the scanty soil and unpredictable climate of Nizhny
Novgorod province, on whose fields the genuine Russian cereal
comes into head. This is why "only vodka from Russia is real
Russian vodka".

Chronology of the Rise and Development of Russian Vodka Production

1386–1398: Genoese merchants first brought grape spirit (*aqua
vitae*) into Russia. The drink became known in the court of the

Great Prince, but made no particular impression. The reaction to it was neutral, like the reaction to other exotic trifles which did not affect Russia in any way.

1426: samples of *aqua vitae* were imported into Russia in greater quantities both from Italy, whence they were brought by Greek monks and Church hierarchs, and from Cafta; in the latter case they were brought by Genoese passing through Moscow on their way to Lithuania. This time the "potion" was deemed harmful, and a ban followed on imports of *aqua vitae* into the Muscovite state.

1448–74: during this period vodka production became established in Russia, with the development of grain spirit distilling based on local raw materials; that is, rye. In the same period a monopoly was introduced not only on the production and sale of "grain wine", but also on all other alcoholic drinks – including mead and beer – which had never before been subject to taxation. From 1474 alcohol production was absolutely reserved for the state.

1470–90: the Great Prince entered into contention with the Church, aiming to prohibit it from producing alcoholic beverages and to close the breach it had made in the state vodka monopoly. Intent on maintaining its privileges, the Church resisted the monopoly.

1505: exports of Russian vodka to neighbouring countries (Estonia, then ruled by the Livonian Order, and Sweden) were recorded for the first time.

1533: the first "Tsar's tavern" was established in Moscow, and trade in vodka was exclusively in the hands of the administration, at least in the principality of Moscow.

1590s: strict instructions were issued to the governors of provinces remote from Moscow to halt all private trade in vodka in alehouses and liquor shops, concentrating it exclusively in the "Tsar's taverns". The tavern-keepers were made responsible for the production and sale of vodka, and the distilling took place in the taverns themselves. The tavern-keepers and their assistants were elected by the community; they were accountable to the governor of the province or region and to the Moscow Offices, the New Quarter and the Office of the Great Palace. That is, they answered to the departments controlling finances and grain supplies, and to the palace administration. They handed over their annual takings, "with an increase compared to previous years", but in other respects were totally free of controls. This system, which lasted until the middle of the seventeenth century, received the name "the sale of drinks on trust"; the tavern-keepers acted both as contractors for the state and as its trusted administrators in the implementation of the vodka monopoly. In the conditions of Russia the production and sale of vodka "on trust" led to large-scale corruption, including bribery and other abuses in the fields of administration and finance, as well as to the spread of theft and drunkenness – in short, to precisely those sad phenomena which to this day are considered "specifically Russian", but which were not characteristic of Russia before the advent of distilling and of vodka.

1648: "tavern revolts" occurred in Moscow and other cities of Russia. These had a multitude of causes: financial abuses by the tavern-keepers, the growth of bribery, a sharp fall in the quality of vodka because of the theft and adulteration of the raw materials, and the ruinous effects on the population of drunkenness, including in several years the widespread failure to sow crops through overindulgence at Easter. A critical factor was the inability of the urban poor to pay their tavern debts. The disturbances spread to include uprisings by peasants in regions near the towns. After the revolts had been put down the Tsar (Aleksey Mikhailovich) called a session of the Assembly of the Land (zemskii sobor), which came to be referred to as the "tavern assembly", since the main matter dealt with was the reform of sales of alcoholic beverages.

1651–2: the system of the "tax farm" (otkup), which had been introduced during a period when the government was extremely short of funds and which involved handing over entire provinces to the power of avaricious vodka concessionaires, was abolished. The sale of vodka for credit, which led people into debt and ultimately to semi-enslavement, was forbidden. Secret, private taverns were closed down. Sermons against drunkenness preached in the churches become more fiery. The tavern liquor-sellers were put under scrutiny, and many were replaced in an effort to rid the trade of openly corrupt elements. The "democratic" election of tavern-keepers from among "honest" citizens was restored, and the system of selling vodka "on trust" was reaffirmed. All this, however, merely acted as a palliative, and by 1659 the situation was as bad as it had been in 1648.

1663: the need of the state for money was such that the farm was partly restored in a number of regions where the sale of vodka "on trust" had not brought increased revenues. Considerations of profitability and free competition between the official tavern system and the liquor shops again led to frightful orgies of drunkenness and to the widespread rise of illicit distilling. Working people were ruined, and their energies, which constituted the primary capital of the state, were dissipated.

1681: the government was forced to reinstitute a strict state monopoly on distilling and on the sale of alcoholic liquor, even though this was not as profitable as the mixed private–state system. Along with this reform the government introduced a new system for the production of vodka: the provision of vodka to the treasury at strictly fixed prices, or as the commodity equivalent of a tax. The people who acted as contractors had to be members of the gentry, owners of large estates, and they had to give written undertakings that they would furnish specified quantities of vodka to the treasury within a given time. The adoption of such a system was unavoidable, since the government did not possess its own distilleries. The vodka delivered to the state was held in government stores manned by armed military guards responsible directly to the Tsar. To receive and distribute these supplies of vodka special officials were appointed – the *burmistry*, whose numbers were kept relatively small so that control could be easily maintained. However, the *burmistry* as well were unable to resist the temptations associated with their duties. Either they took bribes from landowners who had not delivered the agreed quantities of vodka, or they cheated the vodka retailers, trading on their own account or squandering the stock of vodka in their stores.

1705: Peter I decided firmly that the main priority for the state during the Northern War had to be to obtain the greatest possible profit from the sale of vodka; and moreover, that it was necessary to obtain payment in advance rather than collecting this piecemeal following retail vodka sales. He therefore reintroduced the farm system throughout the territory of Russia, combining it with state sales and granting the farm concessions to more wealthy, energetic and ruthless individuals. The Tsar calculated that the concessionaires would collect greater revenues, and that even if they did not, the money they put up in advance in farm payments would at least provide him sooner with the money he needed to prosecute the war and equip his fleet. However, this system was retained for only ten years; sensing that the people would not tolerate it any longer, in 1716 Peter introduced freedom of vodka production, without levying a duty on the distillers, on the equipment, or on the finished product. Vodka production was promptly transformed into a branch of agriculture, as large numbers of grain producers took it up.

1765: the government of Catherine II made distilling a privilege of the gentry. It was free of any tax obligations, but the scale on which it could be conducted was limited in accordance with the social status of the landowner involved. Thus princes, counts and the titled nobility were allowed to produce more vodka than the lesser landowning gentry, a system which corresponded to their actual economic capacities. The privilege of distilling and the scale of production permitted were also linked closely with the rank held by the gentleman-distiller in the civil service, indirectly encouraging the gentry to serve the state. Simultaneously, the other orders of society – the clergy, the merchants, the urban masses and the peasants – lost their right to distil spirits and were forced to buy vodka produced in state distilleries. As a result of this system the domestic distilling of the gentry attained a high level of development in both the technology employed and the quality of the product. The vodka of the gentry did not compete with the vodka of the government and did not exert any influence on it; the two coexisted peacefully, since the private product was meant to satisfy the domestic needs of the landowning class. The gentry's vodka did not exert any perceptible pressure on the market for vodka, which was under total state control; the state aimed its production at all social strata except the landowners. Freed from competition, the state kept its product at a medium level of quality, ensuring that revenues would flow into the treasury and that there would be no danger of financial loss. The arrangement enabled the state vodka organization to rest on its laurels, spared the stresses of competitive struggle.

1781: the development of this system led to the founding of official "chambers of beverages" (*piteinye palaty*), which were obliged to deliver set quantities of vodka per year in particular regions, on the basis of the established demand in each locality. The decree did not specify how the chambers were to organize this annual supply of vodka. They could order it from the state factories or, if the state factories could not supply a given region, buy it wherever they chose. The chambers began increasingly to seek non-state contractors who would undertake to produce the vodka required. More and more of these contracts went to friends and acquaintances of the members of the chambers – that is, to people intent on enriching themselves through the treasury orders. This led not only to a renewal of bribery and corruption, but to the gradual reinstitution of the contract-farm system. The government distilleries progressively curtailed their activity as the state's own chambers gave them fewer and fewer orders.

1795: by this time production of vodka by the state had practically ceased, and the only remaining system of government control was the farm. In regions such as Siberia, remote from the centres of state power, the farm system had effectively been in force since 1767, and by the end of the century it was making inroads in St Petersburg itself. The state authorities ignored this phenomenon because of the flourishing domestic vodka production of the gentry. The landowners and the court had ample supplies of vodka; in addition, landowners supplied vodka to their servants and to the peasants, not wishing to see their money go to others. The merchants, through setting themselves up as official spirits contractors, also supplied themselves with vodka. The only social elements which were not directly or indirectly producing vodka were the urban masses, whose needs were met by the state product. In short, the market for vodka was fully satisfied, and so no questions were asked about the sources of supply. State officials turned a blind eye to the breach of regulations and to the loss of revenue that inevitably resulted.

1796: although the government of Catherine II had never entered into conflict with the landowners, when Paul I succeeded to the throne he resolved to bring order to the system and to protect the interests of the state. Thus he aroused the indignation of the gentry and was murdered in 1801. This put paid to any thoughts his son Alexander I might have had of meddling with this delicate issue, bound up as it was with the privileges of the gentry and with the rights of the growing class of Russian merchants. The latter had quietly taken effective control of the state vodka monopoly through the farm system, and had turned it into a source of continuous, uncontested profit. Thus, thanks to vodka, the Russian bourgeoisie at the very dawn of its existence became accustomed not to active competition, but to parasitism and to enriching itself through theft and the adulteration of products, for which the vodka farm offered abundant opportunities.

1819: it was only in this year, following the ruinous war against Napoleon and inflation both in the gold ruble and in the paper currency, that the government of Alexander I finally turned its attention to the farm system. A strict state monopoly was introduced, the first since the time of Boris Godunov. The only exception made was in remote areas of Siberia, where a struggle against the abuses of the vodka concessionaires was still beyond the power of the government. The state took exclusive control of the production of vodka and of wholesale trade, while retail trade was left in private hands. Lacking retail outlets, the state could not introduce a complete monopoly in this area even in the nineteenth century. To prevent speculation in government vodka, a firm price was decreed, to apply throughout the empire – seven paper rubles per bucket. The introduction of the monopoly quickly filled the state coffers. In the space of a year vodka revenues increased from 14 to 23 million rubles. This increase corresponded to the annual sum of which the farm concessionaires had cheated the government; in the period between 1801 and 1820 they had failed to pay the treasury almost 200 million rubles. The retail vodka sellers did their best to exact revenge. They began to cheat on their payments to the state, stealing and adulterating the product, and ultimately brought their contributions to the treasury down to 12 million rubles in 1826. Overall demand for vodka in Russia began to decline, since the domestic distilling by the gentry limited both the spread of drunkenness and the demand for the lower-quality state-produced vodka. The market was saturated, and it was the producer, not the consumer, who suffered. But since in Russia no producer was prepared to earn profits by raising the quality of the product and

engaging in honest competition, the main producers of vodka –
that is, the gentry – demanded that the competition provided by
the state should be abolished.

1826: after the crushing of the Decembrist revolt the new Tsar
Nicholas I, anxious to make a conciliatory gesture to the gentry
and to strengthen the monarchy's position, restored the farm system and from 1828 completely abolished the state monopoly on
vodka. It may seem a strange paradox that this harsh ruler, who
strengthened virtually every area of the state system and the
administration, should have acted in this way. But in reality
there was nothing contradictory about this step. The rulers who
abolished the monopoly on vodka in order to strengthen their
position within the state were not the weakest Russian monarchs, but the strongest: Peter I, Catherine II and Nicholas I. For
these rulers the most important consideration was not profits but
political gain. They were great politicians, not merchants or gro-
cers, and were quite ready to trade off vodka in exchange for
political stability. The price they paid, however, was ruinous not
only for the state treasury but also, in a very direct fashion, for
the people. This was true both in terms of the population's sol-
vency and, to an even greater degree, in their health and morale.
The farm system, which brought gigantic profits to a handful of
avaricious scoundrels, was always hated, since it led to poverty
and the unrestrained growth of drunkenness, and simultaneously
to a decline in the quality of vodka and a corresponding increase
in its destructive effects on the health of the population. The farm
system introduced under Nicholas in 1826 was no exception. In
the space of a quarter-century, by 1851, all its harmful effects
had manifested themselves. These proved especially pernicious in
the suffocating, reactionary atmosphere of the time, when any
criticism was smothered, and the farm system was maintained
through the suppression of the population at bayonet-point. The
common people tried to avoid having to drink the farm vodka.
By the middle of the nineteenth century the profits made from
supplying vodka had gone into a steady decline.

1850s–60s: the government of Nicholas I found itself in dire
need of funds after crushing the revolution of 1848–9 in
Hungary. This period also saw the state facing large outlays
through the beginning of extensive railway construction. Other

major areas of government spending were preparations for the Crimean War and the replacement of the obsolete sailing ships of the Baltic fleet. Meanwhile, in various regions of Russia between 1847 and 1851 a transition was being gradually carried out to an excise-farm system, under which the state held a monopoly of vodka production in its official distilleries and sold vodka at high prices to the farm concessionaires. The government also hoped to derive extra profit from the income derived by the concessionaires from retail trade. The concessionaires, naturally, strove to enrich themselves by extracting from their customers not only the retail margin for the treasury but also extra income for themselves. The system was heavily abused and evoked strong popular dissatisfaction. Hence, immediately after the abolition of serfdom in 1861, and as part of the general move towards social and economic reform, the farm system was abolished.

1863: the farm was replaced by an excise system.[5] It is significant that this change had to be introduced progressively over almost fifteen years, since the concessionaires had no intention of yielding their positions voluntarily. Even when they were subjected to organized boycotts, with the tavern vodka-sellers and the consumers of vodka united against them, the concessionaires still found ways to create confusion and trick their adversaries. In some cases they provoked or directly organized peasant uprisings, hoping to bring about the use of military force against the rebels. Or else they handed out large quantities of vodka free to the peasants, in order to gain their goodwill or to disrupt the carrying out of seasonal tasks and thus to harm the masses participating in the boycott. In short, the concessionaires used every available method to defend their privileges. The abolition of serfdom, however, opened the way to the development of capitalism in Russia, and forced the government to respond not to the interests and demands of one or another particular class, but to the laws of capitalism, of the market. That is why the regime chose to introduce not a state monopoly but an excise system, a form which was adapted to the capitalist economy and which was in force in the countries of Western Europe to which the rulers of Russia looked as examples. The excise system, however, did not become firmly implanted in Russia. It failed both to be economically effective and to influence social morality for the better, for several reasons. First, it sharply reduced the price of alcoholic spirits, and the treasury's income from this source promptly fell from 100 million rubles to 85 million. Second, the quality of vodka declined no less abruptly, since the producers of vodka were anxious to avoid losing their profits as a result of low prices. The product was often adulterated, and potatoes were substituted for grain as the raw material. The results included mass poisoning, in some cases fatal. Third, drunkenness, which had become less prevalent during the popular struggle against the farm system, once more reached horrifying proportions. This was not reflected so much in the form of an increase in the volume of consumption of vodka, as in the increasing social and medical ill effects of its consumption. Cheap low-grade vodka "for the people", and the lack of control over the new formulas employed by vodka firms, led to a catastrophic growth of alcoholism, with the appearance of large numbers of chronic alcoholics. This was something which had not been observed before the era of capitalism, for all the centuries-old history of drunkenness in Russia. While pure Russian rye vodka is the most benign of all spirits in its long-term effects on the body, adulterated imitations are among the most deleterious.

1868: only five years after the introduction of the excise system, attempts began to be made to reform it, and to correct the social problems it was causing. There were calls for the excise system to be "democratized" and regulated; it was urged that the number of taverns be limited and that control over the sale of vodka be transferred to the community (in effect, to the local council or *zemstvo*). Other demands were for distillers to be subject to election, and for village assemblies to have the right to forbid the sale of vodka in their locality. However, these projects could not hope to rationalize vodka production, sale or consumption on the level of the state as a whole.

1881: the Council of Ministers decided to introduce more substantial changes. The taverns of the past would be replaced by inns which would not only sell decent vodka, but in which drinkers could also obtain food; this would undoubtedly reduce the incidence of drunkenness. Meanwhile, the question of permitting the sale of bottled vodka for consumption off the premises was raised for the first time in Russia. Until 1885 the only such sales were by the whole bucket; bottles were used only for foreign grape wines, which were imported in these containers from abroad. The transition to the sale of bottled vodka was aimed at allowing people to consume the drink in domestic surroundings, where they would not drink so much at a sitting. These plans, however, ran up against the absence in Russia of a developed, large-scale glass industry. For lack of containers, Russian people a hundred years ago were still drinking vodka in exactly the same way as during the Middle Ages – in a tavern, and very often, in servings of no less than a cup (*charka*), of as much as 150 millilitres.

1882: after discussion of the "vodka question" at the local level, most of the Russian provincial officials called decisively for the introduction of a strictly regulated state monopoly on vodka. Fifteen years of the capitalist experiment had yielded nothing except confusion and deteriorating conditions.

1885: on the basis of recommendations by the government, a partial reform of the excise system was introduced. But, as Professor V.A. Lebedev observed, "The tavern disappeared only from the pages of the Liquor Code; it was reborn in the shape of the 'inn'." In effect nothing changed. The belated and extremely limited introduction of trading in bottled vodka (which was confined to St Petersburg and Moscow) did little to improve the situation. The factory worker who bought a quarter bottle of vodka gulped it down while standing in the doorway of the inn – if only in order to be able to return the bottle and get the deposit back. After centuries in which taverns had served vodka without food, the common people lacked any trace of a cultivated approach to the drinking of vodka, and this blocked any reforms aimed at limiting vodka consumption. It became clear that the reforms to the excise system had failed, and that the excise laws were totally unable to regulate the production and consumption of vodka. The scientific community, headed by D.I. Mendeleyev, and also state figures, eminent jurists and the public in general all expressed support for the reintroduction of a state vodka monopoly.

1894-1902: the introduction of the monopoly was planned in detail: it was conceived not as an isolated legislative act but as a fundamental reform, proceeding cautiously in stages and extending successively through the various regions of Russia. It first affected St Petersburg and Moscow, and later the provinces, initially in the European part of the country and then in the Asiatic regions.

The Five Vodka Monopolies

Let us review everything which has been set out earlier. Throughout the whole course of Russian history, for how long has a vodka monopoly been in force?

As a rule, Soviet reference books and historical works ignore the question of the vodka monopoly in Russia and the USSR. The first and second editions of the *Great Soviet Encyclopedia* and the *Soviet Historical Encyclopedia* contain articles on the subject, but these state that a monopoly on vodka existed in Russia "during the seventeenth century and the first half of the eighteenth century", without giving a precise date for its inception. The reference to the seventeenth century as the period when the monopoly began is made on the basis of the fact that in 1649 a new set of laws was instituted, the "Code of Tsar Aleksei Mikhailovich", in which for the first time a whole chapter, the twenty-fifth, was devoted to distilling and to the trade in vodka. This document codified the edicts of the state concerning vodka production, clearly reflecting the impact which vodka had had on society and the state.

However, the first monopoly on vodka had been established during the 1470s, around the years from 1472 to 1474. This monopoly remained in force until 1553, that is, for around eighty years, without changes or concessions. It was only in 1553 that Tsar Ivan IV allowed a few partial exceptions to the monopoly – for members of his special administrative élite, the *oprichnina*, and for the governors of a few outlying provinces. The earlier monopoly was replaced by another type – a monopoly for hire – the farm. There was this difference, however: that what was involved was not the usual type of farm, based on a contract, but the letting out of the monopoly "on trust". Apart from these exceptions the official monopoly on vodka was retained in important parts of the state until 1605, although the farm concessionaires exerted heavy pressure. Under Boris Godunov in the years from 1598 to 1604 the state vodka monopoly was again strengthened.

The monopoly was not formally abolished in the early seventeenth century, but in practice it disappeared during the period of the peasant war of Ivan Bolotnikov and of the Polish-Swedish intervention, that is, between 1605 and 1613. The tsars of the new Romanov dynasty did not really succeed in restoring it until 1652. It was from this year – that is, after the "tavern congress", and not from 1649 as is stated in the Code – that a strict monopoly on vodka was introduced. This was the second monopoly, and it was maintained as long as a strong state apparatus existed. Then at the end of the seventeenth century came a period of disorder marked by disputes over the succession to the throne, over the abolition of the patriarchate, and over the Petrine reforms. Amid the revolts and popular disturbances associated with these developments the monopoly simply lost its force, as happened repeatedly during Russian history, and had already occurred at the beginning of the seventeenth century.

A monopoly on vodka has always been a distinguishing feature of strong, stable regimes and of tranquillity within the state. As soon as something disturbs the orderly course of domestic politics, the state loses control of vodka. And as soon as vodka is torn from the control of the state, all conceivable disorders begin breaking out in domestic politics. Vodka clearly constitutes an effective index of the state of health of society.

It was in order to put an end to instability that Peter I, as soon as he came to exercise undivided power in 1696 or 1697, again introduced a monopoly on vodka for a relatively brief period of some eighteen to twenty years. This monopoly was in force during the Northern War, at least until its culminating stage when the outcome – complete victory by Russia over Sweden – was already clear. From 1716 the monopoly began gradually to be eroded. In all, this third monopoly lasted from 1697 to 1734, when a transition took place to alternative forms of state supervision over vodka distilling. In 1765 Catherine II abolished the monopoly entirely, transferring the right to produce vodka to the gentry as a privilege of their class.

Attempts by Paul I between 1798 and 1800 and by Alexander I between 1819 and 1825 to reintroduce a monopoly were frustrated by the gentry and the merchants. That is why Nicholas I and his successors Alexander II and Alexander III did not try to change the established order in any fundamental way. They retained the farm system, which was relatively trouble-free for the state despite being ruinous for the population. It was only at the threshold of the twentieth century, in the years between 1894 and 1902, that a complete state monopoly on vodka was again introduced. This was the so-called fourth monopoly, which in 1902 came into force throughout the empire, except for Poland and Finland, where local laws applied. The fourth monopoly lasted in formal terms for fifteen years, until 1917. The Soviet government totally forbade the production of vodka from 1917 to 1923. In 1924 the fifth Russian or first Soviet vodka monopoly was introduced. This has continued without interruption ever since.

Between 1986 and 1990 new laws governing the sale of vodka and severe cuts in the quantities produced led to breaches in domestic political order and stability, but these developments did not change the basic character of the vodka monopoly.

The chronology of the various monopolies on vodka thus reads as follows, showing the most significant years in the development of each. Periods when the monopoly was not in full force are shown by dots. The partial and unsuccessful monopoly on production and wholesale, but not on retail, sale which ran from 1819 to 1828 is not listed.

First monopoly	1474–1553 ... 1598–1605
Second monopoly	1652–1663 ... 1681–1689
Third monopoly	1697–1705 ... 1716 ... 1734 ... 1765
Fourth monopoly	1894 ... 1902–1917
Fifth monopoly	1924–

As can be seen, the intervals between the monopolies were relatively brief; the longest was about 130 years in the eighteenth and nineteenth centuries. However, the periods when the monopoly was in full force were also quite short, apart from one of approximately eighty years under the first monopoly during the years from 1474 to 1553. For most of the duration of the monopolies they were either ineffective or combined with some other system, such as the limited farm which ran from 1705 to 1716.

5

Vodka and Ideology

Vodka under the Tsars

Throughout the history of vodka, its production and sale have been closely connected with state finances and with state economic policy as a whole. In the development both of feudal and of bourgeois society, therefore, vodka acquired great importance.

In these societies the significance of vodka as a factor in political life was extremely contradictory. Vodka appeared to be classless, in the sense that as consumers of the drink all social strata showed an equal need and liking for it. In asserting the right to monopolize production and trade in vodka, and in regulating the distribution of the drink throughout society, the state was able to present itself as an impartial apparatus of control, operating independently and above classes. But in reality, vodka was never a neutral factor in a stratified society. While remaining classless in relation to the interests of consumers, vodka was always employed as a weapon by the ruling class – whichever class that was – in its struggles with the others.

Vodka was a powerful and convenient tool for limiting the influence of one or another class. If those who controlled vodka knew how to use it, and used it appropriately, their rule would be stable. But as soon as the state or ruling class let the reins drop and lost control of vodka, political troubles inevitably followed. Problems that had earlier been hidden emerged into the open.

That is why the supplanting of the feudal ruling class by the bourgeoisie, and then the overthrow of the bourgeoisie by the proletariat, were both accompanied by a change in the ownership of vodka. Control of the drink was transferred into the hands of another ruling class hostile to the old one. At the same time, a related change took place in the political use to which vodka was put; that is, the drink's social significance altered. This change of ownership was invariably painful both for the state in general and for a large part of the population. The transfer of vodka in the eighteenth century from the control of an autocratic monarchy into the hands of the privileged gentry was accompanied by a deterioration in the position of the people, by a sharpening of class divisions between the landowners and peasants, and by a wave of large and small peasant uprisings.

In the nineteenth century the transfer of vodka from the control of the gentry to that of the mercantile and industrial middle class was accompanied by the ruination of the peasantry and by the growth of a dispossessed urban working class. Working people were not only impoverished, but effectively enslaved.

Growing popular unrest forced the government of Tsarist Russia to take control of vodka away from the greedy bourgeoisie and to place it directly in the hands of the state. But already it was too late. The fourth vodka monopoly solved nothing.

Vodka after the Revolution

The Revolution of 1917 swept away the old state, and the first workers' and peasants' government outlawed vodka until public order could be restored in the country. This was a bold and well-advised step, and the results were excellent. As early as 1924, a delegation of British trade unionists who came to Russia on a tour of inspection pointed out in their report[1] that nowhere in the country had they observed the phenomenon known worldwide as "the universal Russian drunkenness". They stated that the Bolsheviks had succeeded in implanting a new psychology in the working class: contempt and hatred for drunkenness as an oppressive and degrading influence on workers. The Bolsheviks, said the unionists, had raised the class pride and dignity of the workers almost to aristocratic levels.

While the naive and blinkered idealism of Western socialists who visited the new Soviet Union is well known and must be allowed for, there is no doubt that there was a genuine change for the better. The members of the delegation wrote:

> The success of the Communist movement must undoubtedly be ascribed to the courage of its leaders, who have not been afraid to admit their errors when their theories have turned out to be inapplicable to real life. The basic principle of the whole system is the formation of a state and social order which would be more favourable to the working class and which would open up the same possibilities to all men and women. . . . Although we do not believe that the social system introduced in Russia would on the whole be suited to our country, the members of the delegation are nevertheless firmly convinced that the new social system is bringing the Russian people great advantages both in the way of cultural benefits, and in relation to their free self-determination.[2]

Stressing the change in regard to vodka and its connection with the political changes that had occurred, the trade unionists noted:

> From 1924 the sale of alcoholic beverages of the type of light wine and beer has been permitted. However, the production and sale of vodka with an alcoholic strength of more than 20 per cent is strictly prohibited.[3] Neither in the towns nor in the clubs are there bars serving alcoholic beverages, and these drinks are usually consumed only after dinner in a restaurant or in people's homes. Members of the workers' clubs who are observed in a drunken condition are subjected to severe punishments or to the loss of their privileges. On the street one may at times encounter people who are somewhat merry, but to those who recall pre-Revolutionary Russia it is quite clear that there are virtually no drunkards on the streets or in public places. The illicit production and sale of vodka is treated as a criminal offence.[4]

The above situation applied until 1936, when steps were taken to expand state sales of strong alcoholic beverages, primarily vodka but also including fortified wines such as port and sherry. There was also a significant increase in sales of "Soviet champagne", and also of other, authentically Russian sparkling wines, Donskoe, Tsymlyanskoe and Razdorskoe. This led to a certain increase in the consumption of alcoholic beverages in the years between 1937 and 1940, but the people who were affected were not for the most part those connected with the productive sector of the economy. The increased consumption was primarily on the part of the so-called creative intelligentsia – writers, journalists, theatrical, operatic and ballet performers, painters, sculptors and architects – and their hangers-on.

This phenomenon was linked directly to the political purposes of certain circles which had created a stronghold for themselves within the organs of law enforcement, justice and the secret police. During the repression under Stalin these people aimed at killing several birds with one stone. They sought to undermine the puritan spirit and ideology of the Leninist old guard and of the mass of party members, and to divert the bohemian passions of the intelligentsia into safe channels. With the help of alcohol, they sought to cripple the tongues of potential critics; and at the same time to reinforce repression by advancing the cause of philistine petty-bourgeois elements, who would form a social and political counterweight to the anti-Stalinist opposition inside the party. Through such methods they tried to shatter the political solidarity of Soviet society. These political ploys with alcohol had dire consequences, releasing a genie which proved difficult to force back into the bottle.

Until 1941, nevertheless, the party and the government retained control over the situation. Drunkenness was condemned and punished as a social evil irreconcilable with proletarian ideology. But behind the façade of order sinister forces were at work, gradually demoralizing society and undermining faith in the unity of word and deed of the state leadership.

In this respect, the war of 1941 to 1945 proved a bitter blow. Almost immediately after the victory at Stalingrad in 1943, for the first time in the history of the Red Army, a ration of 100 grams of vodka per day began officially to be issued. It is true that non-drinkers, of whom there were at first a great many, might be issued with sugar and chocolate in place of vodka (this was the case in special units, and in particular for airborne forces). By 1945, however, almost no one availed himself of this substitute. A profound change had occurred in the psychology of people who until this time had regarded drunkenness as something shameful. Suddenly, heavy drinking acquired merit even in the eyes of the troops' immediate superiors, and refusing the ration of spirits came to be viewed as a sign of unreliability and of insubordinate tendencies.

By 1945 virtually the whole active male population of the Soviet Union was in the army; and after demobilization these new attitudes to alcohol and habits in its use, which had been assimilated by tens of millions of people, were sown far and wide throughout the country. A death-blow was thus dealt to all the work which the Communist Party had performed during the twenty years between 1917 and 1937, when disapproval of drunkenness, and indeed of the consumption of alcohol in general, had been essential to being regarded as a true Communist.

After the war, beginning in 1947, a hypocritical silence surrounded the use of vodka. This was a response to the increase in vodka consumption, which had its origins partly in the war – and was most noticeable among the military – and partly in the lack of educational work by the party. Meanwhile, the official opposition to drunkenness was maintained; the abuse of alcohol by party members was considered unacceptable. No one was willing to acknowledge or discuss this demoralization. At the same time, the moral decay continued – secretly and illicitly, but nevertheless encouraged by the same organs of law enforcement and state security.

A clear contradiction also emerged in that the working class, the people involved directly in production, became the country's main consumers of vodka. Workers in heavy industries such as metallurgy and mining, those in the chemical industry, and also those in transport and construction displayed a special need for the drink. During the 1950s and 1960s the price of vodka was kept low – 2 rubles 65 kopeks a bottle – and with the free availability of the drink in state shops, the demand was translated into ever increasing consumption.

Economic losses eventually forced the authorities to conduct a drive against drunkenness at work. But instead of uncovering and analysing the true causes of the problem, this campaign took on an air of dissimulation. Directed not against the causes of drunkenness but against its consequences, the exhortations merely alienated part of the workforce, yielding neither economic nor political results.

Many years earlier Friedrich Engels had observed that the proletariat, at least under conditions of intense capitalist exploitation, experiences a "vital need for vodka".[5] This is explained by a number of factors. First, alcohol quickly relieves emotional and physical stress. Second, through its high calorific value, it can actually supply the drinker with energy (though at a price in health terms); it can be to some extent a substitute for proper food. Third, the need of the proletariat for vodka may be explained by the psychological craving which any individual who leads an unsatisfied and unfulfilled life experiences for a means of escape. Fourth, we may note an understandable need for alcohol in people engaged in dreary tasks whose boredom is alleviated by a desensitizing and sedative drug; and in some cases alcohol may even mitigate the effects of exposure to harmful industrial chemicals. Workers who consume alcohol gain relief,[6] and this leads to the use of alcohol by new categories of people who are not drunkards, but who are simply taking care of their psychological or physical health. These people include women, a fact which inevitably broadens the circle of those who regularly consume vodka.

Of course, in all the cases listed here alcohol would be unnecessary if the needs of the people involved were satisfied in the proper fashion: with good food, with measures to prevent work-related poisoning or to treat it by the use of normal medicines, and through creating decent conditions for the comfort and leisure of workers once they have finished their shift. These simple needs were not supplied to the great majority of the Soviet working class; especially not to transport and construction workers, for whom the state's concern steadily diminished. Instead of taking proper care of workers, the state simply raised their wages, increasing the bonuses for work in dangerous or unpleasant conditions. Large numbers of workers, as a rule young ones living in hostels, lacked the habits of family life and were, moreover, accustomed as a result of their army service to having their bosses take full responsibility for their welfare. As civilians living in big cities such workers, torn from their native village setting, frequently became disoriented. Unused to managing a budget, they wasted their money, ate poorly, and often sank into alcoholism, losing the ability to function as family breadwinners.

Forcing workers to look after their own needs instead of making proper provision for their daily life and leisure was the main cause of the spread of drunkenness, together with the gradual curtailing of political education in the 1960s and 1970s.

It was officially forbidden to bring vodka into the workers' hostels openly during the daytime, in order to consume it after supper. As a result, vodka was brought in secretly and drunk surreptitiously at night, without any food, only a few hours before the drinkers were due to report for work in the factories. All this led gradually to the situation in which, by the mid-1960s and early 1970s, drunkenness on the job became the norm, a commonplace, everyday phenomenon. Alcoholism came to affect workers of diverse categories, including those in textile production, where the workforce consisted mainly of women.

The one type of behaviour that was not condoned, and was more or less kept in check, was open alcoholic excess and mass drunken brawling. Outbursts of such behaviour were broken up and were followed by exemplary punishments through both legal sanctions and, especially in the second half of the 1970s, financial penalties. The charges were raised for time spent compulsorily in "drying-out" clinics, to which people were consigned when found on the streets in a state of profound intoxication, or when the militia was called to hostels to arrest drunken brawlers.

During the 1970s the Soviet working class underwent a general transition to a new mentality in which there was no place for moralistic condemnations of drunkenness. Thanks to the passing of new and less draconian laws, people had come to perceive the abuse of alcohol not as a social but as a personal matter. It was no longer possible to evoke a popular contempt for drunkenness, to create an atmosphere of disdain for drunkards, since heavy drinking had lost its stamp as a grave offence incompatible with the Soviet system or with membership of the working class. Instead, drunkenness had been transformed into a private matter, part of the personal life of the individual, for all its disruptive effect on family members and work colleagues. The fact that drunkenness also occurred in the official milieu and even in party circles – something which had become common knowledge through a multitude of rumours – served to intensify the hypocrisy which surrounded this social evil.

The Anti-Alcoholism Campaign

There remained an important part of Soviet society which was
untouched by drunkenness, which regarded it with contempt,
and which was shocked and dismayed by the inactivity of the
authorities, who under the cover of formal condemnations had
effectively legalized the abuse of alcohol, ceasing to regard it as
irreconcilable with the ideology of a Communist society. Through
their influence the Central Committee of the Communist Party of
the Soviet Union was led to adopt a resolution on the struggle
against alcoholism, in effect firing the first shots of perestroika.
As a painful social issue which at the same time was not considered "political", alcoholism was an astute choice by the party
and the government, whose campaign of action had to start
somewhere. Beginning the fight with an attack on vodka was a
sensational move which at the same time rested solidly on tradition and involved no political risks. Nevertheless, the manner in
which the leadership attempted to solve the problem of drunkenness was completely incorrect. There was no need to wait for
events to unfold, or to possess the gifts of a prophet, to discern this.

The fact that the government and the party finally took a clear
position, openly declaring war on drunkenness, drew an approv-
ing response from the bulk of the population and even, at first,
from drunkards themselves. Women, aware of the connection
between drunkenness and unruly or violent behaviour, were particularly favourable. Moreover, people in general understood that
drinking during working hours had been the cause of innumer-
able accidents.

In reality, however, the government was bent not on waging a
struggle against drunkenness and against society's hypocritical
attitude on this issue, but on attacking vodka itself. This was a
quite different matter, and represented a completely fraudulent
way of posing the question. The government missed its mark,
and the so-called "anti-alcoholism campaign" was discredited.
The hypocrisy which surrounded the question of alcohol abuse,
the divergence between words and deeds, was not eliminated but
intensified.

The prime need was to prove to society that the government's
words would be matched by action, that the state would

consistently oppose drunkenness as something irreconcilable with socialism and with the ideals of socialist society. But nothing of the kind was done: the task was simply shelved. It was thought that the simple administrative banning of vodka – a completely ineffective measure – would suffice, so no other steps were taken to educate the public or to help drunks overcome their addiction. No serious steps were taken to deal with alcoholism as a manifest symptom of social demoralization.

Today, the fruits of this irresponsibility are apparent to even the most casual observer. Fuelled by the violent, chronic intoxication induced by low-quality illicit liquor, the moral dissipation of Soviet society has deepened. Cynical and mercenary attitudes, born among drunken officials and retailed by profiteering "co-operative" entrepreneurs, have taken the place of proletarian consciousness in the thinking of much of the working class. Only prompt action, aimed directly at the source of the danger, can now prevent the degeneration from becoming complete.

Chronic drunkards should be referred to detoxification centres, either on their own initiative, or on that of their families or of the courts. The spectacle of drunkenness in the streets has to be eliminated; fines, applied to all offenders without exception, need to be imposed for drunken violence, brawling and breaches of the peace. Drunk drivers should forfeit their driving licences for long periods. Those repeatedly convicted of offences deriving from their drunkenness should be referred to specialist help. Since we are now living through a veritable epidemic of drunkenness we should envisage a series of special emergency measures – while never forgetting that drunkenness is only the index of social demoralization, not its first cause. Such measures might include, as a temporary expedient, the introduction of ration cards for spirituous liquors, and especially vodka, on the basis of reasonable social norms. A distinction should be made between blue-collar and white-collar workers, with preference for the former. Also, strict and exemplary measures should be taken to wipe out alcohol abuse among people in positions of authority. Administrators and the heads of enterprises and departments who are implicated in drunkenness should be sacked. Only thus can an example be given to ordinary people.

Instead of these vital steps, the government's key measures in

its anti-alcoholism campaign were directed against the productive
base of the vodka industry, and against liquor production as a
branch of the economy. This was not only monstrously unjust in
itself, but inexplicable on the part of people educated in the
Marxist spirit. The genuine Leninist position – "Production is
always necessary!"[7] – was completely forgotten, and faith was
put in the discredited nostrum of prohibition. The supposedly proletarian state resorted to the idealistic fantasies of the bourgeois
US legislators of the 1920s.

In the course of the anti-alcoholism campaign, wine-producing
state and collective farms were dissolved, thousands of hectares
of vines were uprooted, wineries were shut down or assigned to
other tasks, and the equipment of vodka distilleries was disman-
tled. All these measures were pursued in a spirit reminiscent of
the dawn of industrial capitalism in Britain, when illiterate and
impoverished Luddites smashed the machines in the factories,
thinking that it was the machines that were stealing their bread
and putting them out of work. The fact that at the end of the
twentieth century the leadership of a socialist state could act in
such a fashion testifies to the complete incompetence of this leadership, to its repudiation of the economic and political principles
that should underlie a socialist society.

That is not to mention the losses incurred. The Soviet treasury
lost the very large revenues previously accruing to the state
vodka monopoly. The clumsy prohibitionist measures taken
against vodka were not only utterly ineffective, as could have
been predicted, but contributed to the disruption of state finances
and encouraged the hoarding of sugar for home brewing and distilling. A crucial sector of the economy was lost to potentially
responsible public policy and was made over to burgeoning new
mafias which vigorously undermined socialism.

Other grave errors were committed by the government in its
anti-alcoholism campaign. Absolutely nothing was done to impart
to the population the most elementary teaching on how alcoholic
beverages should be consumed, or even to mount propaganda to
this end. As in previous centuries, the mass of the population
drank alcohol not sitting at a table, but holding on to a lamp-
post; yet the government's resolution against alcohol said not a
word about this problem. This central obstacle to dealing with

drunkenness in Russia was dismissed as "minor" or "unimportant".

No less pressing was the need to break Russians of their national custom of drinking first and eating afterwards. Alcohol should only be drunk after the person involved has eaten his or her fill. In the entire anti-alcohol campaign, not a syllable was uttered about this; nor was there a hint of the importance of this vital matter in the culture of alcohol consumption.

Just as incompetent were the attempts to convince drinkers to substitute "light" alcoholic beverages for vodka, on the basis that by comparison with vodka these drinks were "harmless". These attempts revealed the incompetence of the Soviet state organs, which discredited themselves before the world, since they could not even manage to familiarize themselves with the scientific literature on the topic before issuing their "epoch-making resolution". As is well known, beer and "light" wines, that is, ordinary semi-dry and semi-sweet varieties, not to speak of the fortified wines known to Russians by the contemptuous name of *bormo-tukha*, induce a more serious and chronic form of drunkenness than well purified strong liquors. By comparison with good "Moscow Special" vodka, beer and wine are pure poison. That is why propaganda on behalf of beer and wine amounts in effect to propaganda for alcoholism.

Vodka as a Positive Influence

The way to end drunkenness is through raising the quality of vodka by achieving the greatest possible purification. The price must be raised accordingly – but not arbitrarily, only in line with increases in quality. Scientists, both within Russia and abroad, have always understood these points. Carl Linnaeus, Mikhail Lomonosov, Tobias Lovits, Friedrich Engels, Dmitry Mendeleyev, Academician N.D. Zelinsky – all these distinguished figures have stated one and the same truth: if the country's rulers really care for the people and want to protect them from the dangers of drunkenness, the state must strive to produce vodka which is of the highest possible quality and purity. This vodka will be expensive, but it will not be injurious. The quantity drunk must be limited not by cutting the production of vodka, but by raising the cultural awareness of the vodka drinker.

Society must educate people in a cultured manner of drinking, while punishing recalcitrant elbow-benders and persuading them to take treatment for their affliction. This is the essence of a programme for a struggle against drunkenness that is fitting for all countries, peoples and times. The success of such a program depends on the faithfulness of the authorities to these principles.

What has been outlined here is not an instant cure, and it cannot be implemented over a mere one or two years. It needs to be put into practice patiently, consistently and methodically over decades or longer: only then will it yield results. The prevailing attitude in society must be one of contempt for drunkenness; but at the same time a regard for the high quality of vodka must be maintained. Thus the nature of society's relationship with vodka acquires real significance. The course of history shows that in all countries, and especially in Russia, the least satisfactory periods have been those when society declared the production and consumption of vodka to be an exclusively private matter. Experience demonstrates that the state has no right to take an unclear position here, much less an indifferent, ambiguous or hypocritical one. Such lapses are quickly reflected in the growth of popular cynicism, and hence in the spread of drunkenness.

Capitalist society adopts an equally cynical position in this respect, since it cares nothing for the social consequences of vodka production; its sole concern is profit. Periodically it will resort to an unavailing prohibitionism, with results that are equally calamitous. It is no mystery that capitalist society is now ravaged by drug dependency, with its associated crime and social demoralization, all of which yields such benefit to ruthless capitalists, demagogic politicians and empire-building police chiefs.

Any truly socialist society of the future will possess all the prerequisites for a clear, honest and unswerving line on the question of vodka. But even today the authorities should declare unequivocally that they condemn drunkenness as a matter of principle, and that they will punish, without special favour, those in the state administration who undermine this position. Meanwhile, Marxists are fully aware that vodka, which arose in a historically determined fashion, can pass away only in a natural historical manner, when the necessary conditions for its demise are firmly in place. To conduct a struggle against vodka by legislative means or through arbitrary decisions to curtail production is illogical, obtuse, and incompatible with the Marxist understanding of history and its mechanisms. Marxists observe that for historical and physiological reasons vodka has become indispensable to various categories of industrial workers. Accordingly, they call for a thoughtful and cultivated use of vodka. They demand that its quality should be the highest possible, and that its consumption should be undertaken rationally, to avoid injury to the human organism.

There can be no better illustration of the depths of philistine cynicism to which some high Soviet officials had sunk towards the end of the period of so-called perestroika than the way the leaders of the "State Committee on the Emergency Situation" behaved during their farcical coup of August 1991. According to well authenticated reports, these "defenders of socialism" spent the latter period of their brief reign hopelessly drunk. It is to be hoped that those who subsequently aspire to rule our country will learn from this sad example, and will themselves encourage and practise a cultivated approach to the consumption of vodka.

Public authorities must maintain a tireless effort to explain and demonstrate in practice to every member of society the need for an enlightened approach to alcohol. Everyone must understand that as a product influencing the social, cultural and economic relations of a people, alcohol must be subject to strict controls. If that people cannot provide these controls of its own accord, society will not sit by with its arms folded, but will impose prompt and judicious sanctions on anyone who forgets his or her social obligations. Voluntary self-help organizations to combat alcoholism will be encouraged and supported by firm social discouragement of alcohol abuse.

To sum up, a wide assortment of high-quality vodkas should be placed on open sale at reasonable prices. Propaganda must be disseminated to promote cultivated drinking habits, and the state must inculcate a clear understanding of vodka, assigning to it a definite place in the social system as a product subject to appropriate control. Drunkards requesting treatment should be supplied with every facility to overcome their dependency. It would be good to think that some day there will be a generation of children who have never seen a drunken brawl or a wretched alcoholic.

There we have a flexible, realistic and dialectically balanced programme for solving the vodka problem in an authentically democratic state. Given the difficulties and problems currently afflicting our people in the wake of the collapse of perestroika, it will be as badly needed as ever before. If socialists wish to regain their role as leaders of our society they should be at the fore in urging such policies and such understanding. In this way they will be able to help the Russian people recover a sense of their own tradition, an awareness of Russia's contribution to the world, and a vision in which the cultural development of each citizen leads to a developed culture for all.

Appendices

1. The Gastronomic Significance of Vodka, and How It Should Be Consumed

It was no accident that cognac first appeared in France, with its ancient and highly developed viticulture and winemaking. Nor was it by chance that vodka arose in Russia. Each of these strong spirituous liquors was closely linked not only with the natural conditions of these respective countries, but also with their national cuisines. It is inconceivable that one would drink cognac with Russian *zakuski* (hors d'oeuvre), just as it would be impossible to drink vodka with the characteristic dishes of the French table. Each of these drinks has its own proper accompaniment.

It is true that during the twentieth century, especially in the USA, vodka has been used as the basis for cocktails, since the fact that it has virtually no taste or smell allows it to be employed as a neutral alcoholic product. But cocktails are merely a means of getting drunk, not a gastronomic category, and in any case Russians would never abuse vodka in this fashion. If Americans want to act in that way, that is their business.

The correct role for vodka as a table drink is to accompany and to highlight exclusively Russian national dishes. Above all, vodka is the appropriate drink with meat and meat-cereal dishes, with salty and sharp-tasting dishes, and with fish. It makes an ideal partnership with rich boiled beef dishes of the *boeuf-bouilli* type; with suckling pig roasted whole with *kasha* (buckwheat) stuffing; with a side or saddle of lamb, prepared with onions; with *bliny* (pancakes) eaten with butter, sour cream, caviare or salmon; with *pel'meny* (meat dumplings). Vodka also goes notably well with the idiosyncratically Russian *solianki* – thick highly-spiced soups which may be made from beef, or preferably an assortment of meats; from a mixture of various types of fish; or from a mixture of wild mushrooms; and whose seasonings include green and black olives, capers, finely chopped pickled cucumbers, black pepper, salt and sour cream. A *solianka* is consumed before the entrée, and is served very hot, since it contains relatively little liquid and so cools rapidly. When drunk with these dishes vodka smooths their flavour, invigorating the palate and "cutting" the fattiness of rich foods; it also stimulates the digestion.

The main use of vodka in Russian cuisine, however, is as an obligatory accompaniment to *zakuski*. The range of Russian *zakuski* became established during the eighteenth century. This was also the period when Russian domestic distilling, with its rich and diverse assortment of vodkas, reached its zenith. Thus the character of vodka, its purity and its use of aromatic additives, were all adapted to the taste and to the food content of Russian *zakuski*. Vodka and *zakuski* became indivisible both as an idiom of speech and in gastronomic practice. These concepts became rooted in popular consciousness; but over the following two centuries of social hardship their significance became distorted and was lost to much of the Russian population. Vodka remained the indispensable element, while *zakuski* either lost their richness and variety or disappeared altogether.

Vodka thus came to lose its gastronomic role, which earlier had been a principal aspect of its consumption. The lower orders of the population viewed it solely as a means of getting drunk. It was this which led to alcoholism, to the uncultured and dissolute use of this drink. The abuse of vodka has been encouraged by the authorities in one way or another, consciously or unconsciously, both in the pre-Revolutionary epoch and, more recently, in the quarter-century from the mid-1960s. This has been done through ill considered decisions regulating the trade in vodka, so that it has been sold primarily for consumption on premises which do not even serve food. Off-licence sales have been limited, especially in the case of the sale of vodka in small quantities such as quarter-bottles. But the main problem has been the impoverishment of the Russian table, the change in culinary habits, manners and customs.

The point is not only that vodka is the perfect complement to the dishes of the Russian national cuisine, but that certain Russian dishes also have the property of moderating the effects of alcohol – provided that the menu is composed strictly of the foods such as those listed below, and that the vodka is drunk in moderate quantities. During the last twenty-five or thirty years, however, almost all of the characteristic dishes of the Russian cuisine have disappeared from our national menu. The availability to the general population of Russia's national *zakuski*, with which vodka used to be drunk, has become even more restricted.

The dishes that should accompany vodka include the following:

Meat zakuski: pickled pork fat; ham (Tambov *okorok*); jellied pork; cold suckling pig in aspic; cold pig's head; boiled pork or beef tongue; jellied beef; boiled corned beef; cold veal in aspic. Not only vodka, but mustard and radishes are essential with all these *zakuski*, as accompaniments enhancing their gastronomic attractiveness and setting off their taste.

Fish zakuski: herring with sunflower oil and onions (preferably green); pressed (or less satisfactorily, unpressed) black sturgeon caviare; smoked Astrakhan herring; red salmon caviare; pink *sigovaia* caviare; cured fillet of sturgeon; cold smoked sturgeon; hot smoked stellate sturgeon; sturgeon in aspic; fresh-salted White Sea salmon; smoked belly of salmon; fresh-salted Baltic salmon; salted Siberian salmon; salted hunchback salmon; hot smoked hunchback salmon; fresh-salted silver salmon; cold smoked white salmon (or pink salmon); smoked Baikal *omul* salmon; salted and smoked *shemaia*; cold smoked whitefish; pike in aspic; smoked smelt; lake smelt; salted sprats.

Vegetable zakuski: salted cucumbers; sauerkraut; cabbage *provençal* (stuffed with meat and vegetables); marinated Antonov apples; salted watermelon; salted tomatoes; sour stuffed aubergine; salted mushrooms; marinated mushrooms; Russian *vinegret* (beetroot salad); boiled potatoes with salted herrings.

Vodka is the natural and ideal complement to all these *zakuski*, which are also served with butter and boiled potatoes. The only fish dishes with which radishes are eaten are jellied fish and stellate sturgeon; with salted and cold smoked fish radishes are most inappropriate. While all meat *zakuski* can be eaten together with the vegetable types, the mushroom and fish varieties can be eaten only on their own or with vodka.

Earlier it was mentioned that a number of meat and cereal dishes, or cereal dishes eaten together with salt fish (for example, pancakes with salmon), also need to be combined with vodka. But why specifically with vodka? In principle, fatty meat and cereal dishes always combine well with dry spirits. Thus, for example, Scottish haggis combines superbly with Scotch whisky; one might well say that the pair are inseparable. This example illustrates the delicate distinctions of taste between vodka and whisky and other similar strong drinks.

Haggis provides the best possible complement to Scotch whisky. This is because these authentically Scottish gastronomic products were created on the basis of a single national mentality and set of gastronomic perceptions. Haggis, potatoes, turnips and whisky make up a natural whole, but to combine haggis with vodka is impossible, as impossible as combining Russian *bliny* and caviare with whisky. A confirmed drunkard might perhaps manage such a feat; there are, it seems, people capable of eating haggis with cognac! But here we are discussing phenomena quite outside the field of normal taste, from the realms of pathology or barbarism. In corresponding fashion, vodka provides an appropriate accompaniment only for Russian national dishes, the combination bringing out the specific tastes of both food and drink. Drinking vodka with dishes of European cuisine, even though these might be meat, meat and flour, or fish dishes, is simply not done. To do this would be gastronomic savagery, a breach of the most elementary rules of good taste, a profanation both of the vodka and of the food.

Here it is also appropriate to explain how vodka should be drunk – that is, together with food so that it can perform its true gastronomic role instead of serving as a primitive means of quick intoxication.

As we showed in our historical survey, vodka began to be used in the highest social and cultural circles as an indispensable complement to the Russian cuisine only in the eighteenth century, during the years when domestic distilling by the gentry flourished. It was during this period, and in the high aristocratic milieu, that the proper way of drinking vodka at the dinner table came to be defined. Vodka should be served cold, almost frozen, and drunk in small, barely perceptible mouthfuls (the verb used in Russian is *prigubit'*, suggesting that the drink is tasted with the lips). At the same time *zakuski* and *pirogi* (small meat pies) should be taken from the table. Each sip of vodka should be followed by a mouthful of the food, accenting and setting off its taste.

When vodka is drunk in this cultivated fashion it has virtually no intoxicating effect. This mode of drinking, practised in the circles of the educated, Europeanized Russian aristocracy and gentry, was totally unlike the plebeian manner of drinking vodka in the taverns, where people were obliged to consume the drink straight, and were not provided with food to go with it. The cultivated use of vodka also conflicted with the political aims of Peter I, who made vodka and alcoholic beverages in general an instrument of his absolutist rule. Under Peter vodka became a weapon for policing the population, for undermining social morality, for compromising the integrity of political opponents, and for disrupting the unity of the opposition. That is why Peter waged a struggle against the Old Believers in the Orthodox Church, who espoused sobriety as a principle, and why he introduced the "penalty cup". The latter involved the Tsar sentencing an errant nobleman, military leader, or respected political or administrative figure "in place of punishment" to gulp down a cup of vodka or some other potent drink with a volume of 1.2 litres. This penalty, which was inflicted publicly amid the mocking laughter of other courtiers, always had a sorry outcome. The transgressor became drunk and acted in an undignified fashion (the indignity was multiplied by the fact that he was arrayed in courtly attire, bedecked with all his orders and accompanied by his servants, who were obliged to appear with him in the palace); or else he fell senseless, the victim of sometimes fatal alcoholic poisoning; or he would begin to vomit uncontrollably, also discrediting himself; furthermore, his health might be ruined if he was penalized repeatedly in this fashion. These brutal jokes which Peter played on his political adversaries, with the aim of discrediting them and undermining their prestige, were viewed by ordinary people and even by the courtiers as jests by the "merry Tsar". Eventually they served as a model that was imitated by numerous local "tsars".

The culture of drunkenness that arose as a result of drinking vodka without food was thus reinforced by the Tsar's "penalty binges". The result was described in 1788 by the French physician Le Clerc, who in his book on Russia asserted that to gulp down large quantities of vodka without food was the usual, and indeed correct, way to drink Russian vodka – a view that was adhered to, Le Clerc would have us believe, by the Russians themselves.

Since that time, for more than two hundred years, this fable has been repeated by foreigners describing their travels through Russia. It is retailed today by American businessmen, who with their own eyes observe their Russian business partners behaving in such fashion. Meanwhile, no one troubles to explain that this is not a specifically Russian trait, but simply represents vulgar behaviour that can be found in most countries of the world. The partners more often than not turn out to be representatives of the criminal underworld, or *nouveaux riches* who know and understand nothing of Russian national culture.

The great Russian writers dwelt on the fact that it was precisely those standard-bearers of Russian boorishness, the officers and merchants, who more than anyone else had the opportunity to represent Russian society before foreigners. As a rule these people gave proof of their Russian origins by swilling vodka. But as Mikhail Saltykov-Shchedrin stressed, this was grotesque behaviour; those who indulged in it were representatives not of Russian culture, but only of Russian "savage capitalism".

The ability to drink vodka correctly, and to appreciate its genuine gastronomic qualities and its place in the national menu, is a sign that the drinker possesses the best, authentically national qualities of Russian dignity. For understandable reasons, these qualities are far from being innate in everyone or accessible to everyone, just as most people are unable to drink vodka with rich, costly foods. In short, vodka when correctly served and drunk is far from being a plebeian drink for "muzhiks", as

Western "Russia hands" habitually regard it. Vodka is a drink for gentlemen, since only a true gentleman knows how to drink vodka while remaining totally sober.

The best Russians, the country's cultural and political élite, have always possessed these qualities of the gentleman. In the eighteenth and nineteenth centuries such people included Princes Kurakin and Gagarin, Gorkachov and Lobanov-Rostovsky, Counts Sheremetev and Vorontsov – philosophers and historians who were well known in the West. Also in this category have been such great figures of Russian science and culture as Lomonosov, Lobachevsky, Karamzin, Griboyedov, Herzen, Mendeleyev, Turgenev, Tyutchev, Bunin and Shostakovich. These people loved and valued vodka, and drank it properly. Can anyone say that they were drunks rather than true and cultured gentlemen?

The truly Russian way of drinking vodka should be judged as it was practised by these people, not by the touts, taxi-drivers and black marketeers who turn up beside American businessmen in the bar of the Hotel Intourist. In the same way, we Russians ought to judge the habits of British gentlemen on the basis of the way they behave in their own homes, not by the antics of drunks in sleazy pubs.

Unfortunately, there are not as many true gentlemen in Russia as one would wish, and as a rule these gentlemen sit peacefully at home rather than lounging around outside hotels accosting foreigners in the fashion of today's dissolute youth. That, however, is another question.

To conclude our observations on the gastronomically correct way of drinking vodka, here are a few words on how it should be served. The serving of a drink is of course only a formal thing, but the form should correspond to the content. This point was well expressed at the beginning of this century by the Russian poet Sasha Cherny:

Some will cry: "What's form? A trifle!
If you tip excrement into crystal
Doesn't the crystal lose its shine?"
Others meanwhile answer, "Fools!
The best wine in a chamber-pot

Is not a drink for decent people."
There's no end to this dispute. It's sad ...
But drinking vodka out of crystal –
Now, that's not so bad!

Ideally, vodka should be poured into clear, colourless glass or crystal *stopki*, either cylindrical or in the shape of a truncated cone, with a capacity of 100 millilitres. These vessels should be filled only two-thirds full. To fill glasses to the brim, so that it is difficult to drink from them, is the mark of a philistine. It is a sign of Russian taste, to be sure, but of the lowest provincial and uneducated kind.

That is everything a foreigner needs to know about the correct, and at the same time enjoyable, way of drinking vodka. Anything which does not conform to these rules should be mercilessly rejected as vulgar, uncultivated and historically unauthentic, that is, as the mark of a boor and not of a Russian gentleman.

2. Modern Vodkas of Russia and the Other Republics

The following are the better-known vodkas currently on sale in the CIS:

Moskovskaya Osobaya (Moscow Special)
Russkaya (Russian)
Stolichnaya (Capital City)
Pshenichnaya (wheat vodka)
Limonnaya (lemon vodka)
Posolskaya (Ambassadorial)
Zolotoe Koltso (Golden Ring)
Kubanskaya (Kuban)
Sibirskaya 45% (Siberian 45 per cent)
Yubileynaya 45% (Jubilee 45 per cent)
Krepkaya 56% (strong 56 per cent)
Gorilka
Ukrainskaya s pertsem (Ukrainian pepper vodka)
Pertsovka (pepper vodka)
Zubrovka (buffalo grass vodka)
Extra

Starka (old vodka)
Petrovskaya
– and simply, Vodka

In addition, a number of "vodkas" are produced in the Baltic republics, using different technology and distilled water. These are:

Kristal Dzidrais (Riga, Latvia)
Viru Valge (Tallinn, Estonia)

Of the vodkas listed above Sibirskaya, Yubileynaya and Krepkaya have a strength greater than 40 per cent, and strictly speaking are outside the classical definition of vodka. The hallmarks of a true vodka are a content of pure spirit of 40 per cent, calculated by weight and not volume; the use of rye malt and rye grain (with the addition of other cereals in permitted quantities); the use of soft water from rivers of the Moscow region (2–4 milligrams equivalent of dissolved substances), and the absence of any other ingredients altering the classic taste. The only vodka which fully meets these criteria is Moskovskaya Osobaya. Stolichnaya is prepared with the use of added sugar (in tiny, allegedly undetectable quantities); this is supposed to add to the drink's "velvety" qualities. Limonnaya, Zubrovka, Pertsovka, Starka and Petrovskaya also have flavourings, giving them a particular nuance and distinguishing them from the real Moscow vodka. Gorilka and Ukrainskaya s pertsem, apart from incorporating a series of aromatic additives, are prepared from wheat and from wheat malt, and include quantities of the potato spirit which is traditional for Ukrainian "Cherkassk" vodka.

Pshenichnaya is based not on rye, but entirely on wheat, and consequently also differs from real Moscow vodka. Finally, Russkaya undergoes multiple distillation with a small quantity of cinnamon, which does nothing to improve the drink's taste and quality. Russkaya is an unsuccessful modern brand first produced in the 1970s, and the addition of cinnamon is evidently dictated by the need to mask the potato spirit included in the formula. The name, with its suggestion of authenticity, is confusing for the consumer. Russkaya also contains distilled water, not "living" water from a stream.

The brands which come closest to Moskovskaya Osobaya in terms of the criteria listed earlier are Posolskaya and Zolotoe Koltso; but these are relatively new brands, and full details of where and how they are made have not been published. Zolotoe Koltso uses water not from Mytishchi but from the Klyazma river, so one cannot state with complete certainty that these brands are truly comparable with Moskovskaya Osobaya. Nevertheless, they are of high quality.

The modifications employed in some of the vodkas listed above are traditional for the Russian and Soviet vodka industry. They reflect particular purposes for which vodka is used (for example, Stolichnaya is well suited to cocktails), or cater to specific tastes (as in the case of Limonnaya, Pertsovka and Starka). But if one proceeds from the strict gastronomic principles governing the harmonious use of vodka to complement the dishes of Russian cuisine and its unique *zakuski*, the only vodka which really suits them is Moskovskaya Osobaya. Limonnaya, whose pleasant flavouring has given it a reputation as a "women's vodka", is suitable to a degree for cold fish *zakuski*.

By the way, Sibirskaya should not be confused with a pseudo-vodka produced in Italy, the so-called Siberian Lemon Special, which is nothing like a vodka of the classic type. It is in fact a bitter tincture, standing closer to the *erofeichi* than to vodkas.

3. The Alcoholic Strength of Wines and Spirits

The quantity of alcohol in beverages is usually expressed in terms of the percentage by volume or by weight. The use of different systems for expressing the composition of alcoholic drinks is to be explained both historically – that is, as a result of the practice applying in the country where the drink was produced or in the firm that prepared it – and as a reflection of the technological level of the productive process. Thus, for example, the alcoholic content of vodka used to be expressed exclusively in terms of volume. This was also reflected in the technical description of vodka as "double-tested", "triple-tested", "quadruple-tested" and so forth. But since D.I. Mendeleyev reformed vodka blending, and showed that the combining of grain spirit with water should be done not on the basis of simply adding one volume to another, but by the precise weighing of a definite proportion of spirit, the alcoholic strength of vodka has been expressed as a proportion by weight.

In Russia the alcoholic content of wines and spirits is usually shown on the labels in "degrees", representing percentage by weight and symbolized by the usual degree sign: 40°, 16°. However, this practice is not universal in the CIS. At times, especially in other republics, producers express alcoholic strength with a percentage symbol: "Selected Cognac, Ararat, Armenian SSR. Bottled in the Moscow Wine and Cognac Plant. Strength 42%"; "Ukrainian Pepper Liqueur. Strength 40%"; "Bitter Lemon Liqueur. Kristal Bottling Plant, Moscow. Strength 40%". This manner of indicating alcoholic content in percentages has been used in various foreign countries, and was adopted in a number of republics of the former USSR through the influence of people returning from abroad to work in the liquor industry, or in the design of labels. Naturally, this violation of the established practice in our country is confusing for buyers and must be viewed as an example of technical and commercial illiteracy, deserving appropriate sanctions.

Indicating alcoholic content in terms of a percentage may not tell us whether it was calculated on the basis of weight or of volume. In countries where the production of alcoholic liquors is entrusted entirely to private firms, some producers think it advisable to show some percentage or other as a rough way of allowing the consumer to get his or her bearings (for example, the Finnish liqueur Poliar, 29%). Other producers consider such information unnecessary, especially in the case of wines and liqueurs, every one of which has its individual character. The merits of these drinks are to be found in their bouquet, not in their alcoholic strength expressed in arithmetical terms. Producers who followed this latter course included, for example, the Portuguese port houses, and also many Italian and southern French firms – though the dead hand of European Community bureaucrats now compels all to mark the strength.

In our country, by contrast, a clear system for expressing the alcoholic content of liquors has been worked out. When measurement is by weight, the percentage is shown in degrees. Drinks described in this way are as a rule strong liquors of high quality – vodkas and cognacs. When the percentages are in terms of volume, they are shown on the labels as follows: "Al Sharab, Table Rose, Baku Wine Plant no. 1, strength 9–14% by volume"; "Matrasa, Table Red, Baku Wine Plant no. 1, strength 10–14% by volume"; "Kagor, no. 32, Kuban Vineyard Industry, spirit content 16% by volume". Other ingredients of the wine are listed on the labels and shown as proportions by weight. Sugar content is shown on labels as "sugar 16%", or "sugar 30–40 g/l" – that is, grams per litre. Knowing the density of alcohol, sugar and other ingredients contained in the drink, one can translate volume measurements into percentages by weight and vice versa. Sometimes the labels also specify the quantity of titrated acids (that is, the quantity determined by chemical analysis), especially in the case of wines that are specially blended, or wine merchants' "house" brands. Thus the label on the white dessert wine Ulybka, from Gelendzhik, specifies: "alcoholic strength 15%, sugar content 14%, titrated acids 5–7 g/l".

Unlike the case with vodka, the alcoholic content of which is always expressed in terms of percentage by weight, the alcoholic content of wines and spirits alike in Western European countries is measured in terms of volume. This is now reckoned after the alcohol and water have been combined, which gives a mixture with a smaller volume than that of the two original constituents; until the middle of the twentieth century it was often reckoned by the volume before mixing. The archaic "proof" measurement still used in the USA, formerly a proportion of an arbitrary maximum legal strength, has been altered so that the figure is now simply double the percentage by volume. The British proof system, similar but working from a different norm, has been dropped. But in the past few years our own exporting organizations, trying to conform to Western norms, have also begun to indicate the alcoholic content of vodka not in degrees, but in percentages (Moscow Special, 40%), without understanding the confusion they are causing or the loss of vodka's distinctive and accurate measurement system.

The volume system now standard in the European Community might be considered practical because of its apparent uniformity. But in fact, it does not correspond linearly to the amount of alcohol in the drink, since the more spirit is added to the water, the more the resulting mixture "shrinks" in relation to the volume of the original liquids. It is therefore the Russian system of calculating the strength of alcoholic liquors that is the more accurate, since it gives the precise content by weight of alcohol in the drink. A "degree" of alcoholic content in the Mendeleyev system constitutes a one-hundredth part by weight of pure spirit, weighing 0.789 grams per millilitre.

4. The Effects of Alcohol on the Human Body

An enormous amount has been written about the effects of alcohol on human health and behaviour – that is, on the way alcohol affects people's nervous systems, their physical co-ordination, and their self-control. This literature is heavily influenced by emotion, and is moralizing in tone. The medical arguments it employs should convince no one, since the cases cited are pathological in nature, drawn from observations of chronic alcoholics. These horrifying examples are in no way characteristic of normal users of alcohol.

It is not surprising that the usual medical propaganda against drunkenness makes no real impact; least of all on the physicians themselves, since it is well known that the medical profession contains more than its share of drunkards. The medical tracts supply no concrete information on how this or that dose of alcohol affects a normal, healthy human being who is neither an alcoholic nor a stranger to alcohol; that is, the kind of person whom Lenin in his time described as "neither a monk nor a barfly".

In 1903 the Russian physiologist N. Volovich, trying to reach an objective understanding of the processes which occur when a normal, strong, healthy male human being consumes alcohol, conducted a unique experiment. Rejecting the normal method of assessing the degree of drunkenness on the basis of the subject's external demeanour (excited, merry, kissing those around him, staggering, clinging to the wall, crawling under the table), Volovich devised a genuinely scientific system of measurement. As his basis for assessing the influence of alcohol on the human organism, he chose an objective phenomenon: the subject's pulse rate after he had consumed various quantities of alcohol, compared with that after drinking a glass of water. The pulse was monitored over a period of twenty-four hours after the subject consumed alcohol. The following results were obtained.

Drinking 20 grams of alcohol (in this case, 96 per cent pure absolute alcohol) produced virtually no changes; over twenty-four hours only 10 or 15 extra beats were registered, or none at all in the healthier subjects. Further up the scale, 30 grams of alcohol produced an extra 430 beats. After that the effects increased as follows: 60 grams of alcohol, 1872 extra beats; 120 grams of alcohol, 12,980 extra beats; 180 grams of alcohol, 18,432 extra beats; 240 grams of alcohol, 23,904 extra beats.

When 240 grams were consumed, an extra 25,488 beats were also registered on the day after the experiment. This represented the residual effects of alcohol consumption, since twenty-four hours were not enough for a large dose to be eliminated from the body.

From these data Volovich drew the conclusion that the consumption of 20 grams of alcohol produces no negative changes in the human organism, serving only in a stimulating and cleansing role. Therefore, consuming such quantities each day should be regarded as entirely normal, and even as necessary for prophylactic reasons during particular periods (autumn, winter, damp weather and so forth). If Volovich's 20 grams are converted into vodka with a spirit content of 40 per cent, we arrive at a daily figure of 50 grams. Thirty grams of spirit, or 75 grams of vodka, can be regarded as the limit for normal use. A daily intake in excess of 60 grams of spirit – that is, 150 grams of vodka – is already harmful, and more than 100 grams of spirit or 250 grams of vodka is simply dangerous, since this signifies an average of 10 to 12 thousand extra heartbeats over twenty-four hours, or from 8 to 10 per minute more than the organism requires. There are many people who are unable to control themselves in such circumstances. Everyone should have access to this information so that they can work out their strengths and capabilities, without entertaining illusions.

Meanwhile, even 30 grams of vodka is a perfectly adequate amount to accompany fatty dishes, or to complement salty or spiced fish or vegetable foods (see the examples in Appendix 1). Such food is not consumed every day, but it is normally eaten at least two or three times a week. Hence a quantity of 100 to 150 grams of vodka per week or 400 to 500 grams per month represents normal consumption. This amount should serve as a basis for calculating the minimum state production of vodka and the personal consumption level at which people should aim. It follows that a country which has at least 200 million potential consumers of vodka should produce and sell no less than 100 million litres of vodka per month to meet normal levels of demand. This figure is a minimum which should be officially recognized. Production cannot be allowed to fall below this without risking an explosive situation as demand goes unsatisfied.

5. The Raw Materials and Production Techniques of Other Principal Spirits of the World

Name of drink and country of origin	Raw materials	Peculiarities of the production process or of the processing of the raw materials which affect quality or flavour
Cognac (France)	Grapes	Simply but carefully distilled from wine. Matured for many years in oak casks. Other countries make similar brandies; some, such as Armenian brandy, can reach a high standard.
Marc (France)	Grape skins and pips	A large group of rough spirits made from waste materials from winemaking; another example is Italian grappa. Raisiny taste.
Genever (Netherlands)	Barley malt and wheat	Multiple distillation, with addition of juniper etc.
Gin (England)	Barley	Multiple distillation, with addition of juniper berries and other flavourings.

Name of drink and country of origin	Raw materials	Peculiarities of the production process or of the processing of the raw materials which affect quality or flavour
Whisky (Scotland)	Barley, barley malt in various proportions	The grain is first moistened so that it germinates, then dried in a stream of hot smoke from a peat fire, partly roasting it. The spirit is matured in wooden casks for at least five years. Most brands are blends with neutral grain spirit.
Whiskey (Ireland)	Barley malt, with rye, barley and maize	The grain is first moistened, then dried in the open air until hard. The spirit is matured in oak casks for five years.
Bourbon (USA)	Principally maize, with wheat and wheat malt	The spirit is matured in barrels whose interior has been scorched. This imparts a dark cognac-like colour and masks the maize odour.
Canadian whisky	Rye and wheat grain, potato spirit	The raw grain spirit is blended with pure ethyl alcohol and potato spirit, then subjected to secondary distillation. The alcoholic strength is always 43°.
Schnapps (Germany)	Potatoes, beet, wheat malt, barley	Many types are flavoured with herbs or spices.
Sake (Japan)	Rice	The rice is prepared by a special process involving strong steam heating of the grain before distillation. *Sake* is colourless. It is drunk warm, in tiny quantities.

Name of drink and country of origin	Raw materials	Peculiarities of the production process or of the processing of the raw materials which affect quality or flavour
Mao tai (China)	Cracked rice, rice malt	Yellowish in colour, with a distinctive odour.
Khan shi na (China)	Sorghum	Similar to *mao tai*, but cloudy and of inferior quality.
Bambuse (Indonesia)	Bamboo seed, from several high-yielding varieties. Bamboo flowers once every 25 to 30 years.	Despite multiple distillation, *bambuse* is poorly cleansed of contaminants, especially methyl alcohol. It causes hallucinations, and is therefore used only very rarely, on special feast days, as a ritual drink, by local religious cults.
Rum (Latin America, West Indies)	Sugar cane stems and pulp; products of sugar manufacture, including cane juice and molasses	The different types of raw materials result in a wide range of rums, differing in aroma, alcohol content and general quality. Strength is up to 55°.
Calvados (Normandy)	Apples, in good condition, not overripe	Matured in casks. Alcoholic strength varies from 38° to 50°, depending on the type of raw material and the distillation process.
Kirschwasser (South and south-western Germany, Switzerland)	Fermented mash of small black cherries, together with stones	A dry white spirit, not to be confused with sweet cherry brandy. Stones give it an almond flavour.

Name of drink and country of origin	Raw materials	Peculiarities of the production process or of the processing of the raw materials which affect quality or flavour
Pejsachowa (Ukraine, Israel)	Raisins	Produced by double and triple distillation.
Slivovitz (Hungary, Czechoslovakia, Romania, Yugoslavia)	Greengages	Produced by drying of the fruit before fermentation, multiple distillation.
Tutovka (Azerbaijan, Armenia)	White and black mulberries	Often produced by multiple distillation. Yellowish-green tinge, unusual aroma.
Kizlyarka (Northern Caucasus, Stavropol, Kuban)	Various fruits, including apples, pears, plums, apricots	Variable, depending on raw material and quality of preparation.
Veynovaya vodka (Southern republics of CIS)	Spoiled, sour grape wine or grape vinegar, with the addition of low-quality grapes (sometimes unripe)	Variable quality, often very rough.
Kumyshka (Udmurtia, Mariy El, Bashkiria)	Fermented cow's and sheep's milk, *kumys* (fermented mare's milk)	A rough, home-made product.
Arka (Kalmykia, Buryatia)	*Kumys*; or fermented cow's milk with the addition of *kumys*	As *kumyshka*, but *arka* is always drunk hot, since when cold it has an unpleasant odour.

Name of drink and country of origin	Raw materials	Peculiarities of the production process or of the processing of the raw materials which affect quality or flavour
Arza (khorza) (Kalmykia)	*Arka*, or *arka* with *kumys*	Produced by multiple distillation of *arka*. It is a very strong "milk vodka" which is always drunk hot.
Pulque (Mexico)	Agave sap	Strength of 32–34°. The quality of the drink varies widely, depending on the quality of the raw material, the technology of production and the degree of purification. There is virtually no established standard.
Raki (Turkey)	Dates	Aniseed flavoured; Greek *ouzo* is similar.
Arak (Indonesia, Malaysia, Thailand)	Mixed ingredients: rice, palm juice, cane molasses, coconuts	Similar to rum. The best types are Javanese. Its strength is usually 58°.

Afterword

We have explored the history of the origins and evolution of vodka as a product which was developed in Russia and whose role, though extremely contradictory, has by no means been minor. Despite the fragmentary nature of this account, and the specialized character of chapters devoted to specific problems, the reader should have formed a broad impression of the history of vodka. In any case, he or she will have perused a good deal of factual material which should have dispelled any notion that vodka as a topic does not deserve the attention and researches of historians.

It is to be hoped that the present work will stimulate further research into the secrets which still surround vodka, in the fields both of the physical sciences and of history. The discovery and study of previously unknown archaeological and written historical materials could serve to illuminate the most obscure period in the history of vodka, the period of its genesis – that is, the end of the fourteenth century and the first half of the fifteenth. Such research could elucidate the social, political and economic significance of vodka in the life of the state, revealing vodka as a factor which cannot be ruled out in any scholarly study of social processes.

All such work helps us to gain a more precise understanding of the true history of our country and of the Russian people, and will also suggest when and how we will succeed in ending the status of vodka as a symbol of social evil. Once this has been achieved, people will be able to consume vodka in a proper, civilized fashion for gastronomic pleasure and for the sake of their own health and serenity, since our Russian vodka from a scientific point of view is the purest and most benign of all alcoholic drinks in the world.

It is only necessary to ensure, through legislation and through our conscientious labour, that all the vodka produced in our country is prepared exclusively from the true and traditional materials, rye and pure stream water, and that it is made and purified to the highest possible standard.

A utopian dream? No, it can be done.

W. Pokhlebkin

Notes

Chapter 1

1. "Fiscal" is here understood as relating to the effect of the sale of alcohol on the state budget and finances; "productive" in terms of acquiring and processing agricultural raw materials, creating and technically equipping the vodka industry, and incorporating it into the general economic system of the country; and "social" in relation to the development of a culture adapted to alcohol, and the resultant widely tolerated drunkenness causing loss of labour time, deterioration in personal and family relationships, and a burden on health services and the forces of public order.

2. P. Alekseev, *Dopolnenie k tserkovnomu slovariu*, vol. 1, Moscow 1773, p. 154.

3. See A.G. Preobrazhenskii, *Etimologicheskii slovar' russkogo iazyka*, Moscow 1959, vol. 1; also M. Fasmer, *Etimologicheskii slovar' russkogo iazyka*, Moscow 1964, vol. 1.

4. V.I. Dal, *Tolkovii slovar' zhivogo russkogo iazyka*, 2nd edition, Moscow 1883, vol. 1, pp. 208, 221–222.

5. F.P. Filin (ed.), *Slovar' russkikh narodnykh govorov*, 13 vols, Moscow 1965–1977. *Vodka* is defined in vol. 4, p. 338.

6. *Slovnik jazyka staroslovenskeho*, 42 vols, Prague 1958–1989.

7. See the Gospel According to St John, ch. 4, v. 2; also P. Alekseev, *Dopolnenie k tserkovnomu slovariu*, Moscow 1776, p. 36.

8. See John, ch. 4; v. 11, ch. 6, v. 55. P. Alekseev, *Tserkovnii slovar'*, 2nd edition, vol. 2, Moscow 1793, p. 298: "And 'drinking water' [*pivnaia voda*] in Latin is *potabilis*."

9. *Voda* was set against *vino* (wine) on the basis of its colour, its effects and its place in religious ritual. The word *vodopitie* (the drinking of water) signified sobriety and abstinence, while *vinopitie* (the drinking of wine) implied drunkenness. See P. Alekseev, *Dopolnenie k tserkovnomu slovariu*, pp. 33, 36; *Sobornii list*, p. 189. See also the use of "wine bibber" (*vinopiitsa*) in the sense of "drunkard" in Matthew 11: 19; P. Alekseev, *Tserkovnii slovar'*, 2nd edn, vol. 1, Moscow 1773, p. 41. Sects which celebrated the sacrament with water instead of wine – something which was considered an extremely grave heresy – were known as *vodianie*, *akvarii* or *idroparastaty* (from the Greek). From this stemmed a profound contempt for any individual who did not drink alcohol; such a person was considered alien and unchristian. Among Russians this age-old attitude is alive to this day.

10. *Slovnik jazyka staroslovenskeho*, vol. 1, p. 190.

11. See P. Alekseev, *Tserkovnii slovar'*, vol. 1, p. 41; *Dopolnenie k tserkovnomu slovariu*, p. 33.

12. In ancient times in Greece, Rome and throughout the entire Hellenistic East wine was never drunk in its pure form. This tradition was maintained in Byzantium, from where the custom was passed on to Russia of drinking wine (and later also vodka) only in a mixture with water. The ancient Greeks and Romans diluted wine in the following fashion: to three parts of water they added one part of wine; or to five parts of water, two parts of wine. A mixture of equal parts of water and wine was considered too strong, a drink only for alcoholics.

13. *Slovnik jazyka staroslovenskeho*, vol. 1, p. 190.

14. Ibid., p. 190.

15. Ibid., p. 633.

16. Ibid., p. 524.

17. R. Brandt (ed.), *Grigorovichev parameinik, v slichenii s drugimi parameinikami*, Moscow 1894–1901. The original, from the twelfth and thirteenth centuries, is stored in the Lenin Library.

18. *Ostromirovo Evangelie.* See *Slovnik jezyka staroslovenskeho*, vol. 18, p. 199.
19. *Suprasl'skaia rukopis'.* See *Slovnik jezyka staroslovenskeho*, vol. 2, p. 200; also vol. 1, p. 120: "Honey drips from the mouth of a wanton woman."
20. Ibid., vol. 18, pp. 199–200.
21. pp. 469–70.
22. I. Zabelin, *Istoriia russkoi zhizni s drevneishikh vremen*, Moscow 1876, vol. 1, p. 468.
23. The Swedish resident in Moscow, Johann de Rodes, in his reports to Queen Christina on the state of Russia from 1650 to 1655, no longer mentioned honey as an object of export, but only wax; although honey was in fact being exported at this time through the Baltic regions. Two decades later, between 1672 and 1674, I.F. Kilburger stated definitely that neither honey nor wax was any longer being exported from Russia. This clearly illustrates both the increase in consumption of these products in Russia itself, and the decline in resources. The traveller Albertus Campensis wrote in 1523 that Muscovy was very rich in honey. At this point honey was still being exported, but local consumption was already falling or, to be more precise, supply had been unable to match demand since the fifteenth century. The decline in the quantity of honey available for use as the raw material for alcoholic beverages was thus already apparent at the end of the fifteenth century, and in the course of the sixteenth century the need became evident to replace honey with some cheaper and more readily available ingredient. Before this time the idea of distilling alcohol from grain could not have arisen simply for historical reasons. Thus the fifteenth century was the apogee of mead brewing, and the beginning of the sixteenth century witnessed the first signs of its decline. See B.G. Kurz, *Sostoianie Rossii v 1650–1665 gg. po doneseniiam Rodesa*, Moscow 1915; B.G. Kurz, *Sochineniia Kilburgera o russkoi torgovle v tsarstvovanie Alekseia Mikhailovicha*, Kiev 1915, pp. 112, 305–6; *Biblioteka inostrannykh pisatelei o Rossii*, vol. 1, 1836, pp. 30–31.
24. *Ostromirov Evangelie* compares *kvas* with *sikera*. See *Slovnik*

jezyka staroslovenskeho, vol. 11, p. 19. Ibid., from the twelfth century: "A bishop should be without vices, neither a drunkard [kvasnik], nor a brawler."

25. Codex zographensis, edn published Berlin 1879.
26. Slovnik jezyka staroslovenskeho, vol. 2, p. 19.
27. Grigorovichev parameinik.
28. Euchologium sinaiticum. Slovnik jezyka staroslovenskeho, vol. 2, p. 19.
29. Ibid., vol. 1, p. 802 (ispl"n') and pp. 800–801. To understand the character of kvas in the Old Russian sense of this word, it is important to note that a specific illness was associated with "unfinished" kvas. This ailment was known as iazia kvasnaia (kvas pain). It affected not the stomach but the head, and resembled the effect which in the seventeenth century was ascribed to hops and called pokhmel'e (from khmel', "hops"). In the sources we find the following references to the iazia: "A prayer for those who have fallen ill through drinking kvas", and "To all those who hated you, you gave such quantities of 'unfinished kvas' that they suffered from 'kvas pain'." In the Russian language between the seventeenth and nineteenth centuries the verb iaziats'ia meant to sway, to be unsteady on one's feet and in one's mind. This verb, with its origins in the noun iazia, signifying a pain incurred through drinking kvas, shows clearly that iazia was a kind of alcoholic poisoning characterized by severe headache, physical weakness, and even a feeling that one might be going mad.
30. Luke 1: 15. See P. Alekseev, Tserkovnii slovar', vol. 3, St Petersburg 1794, p. 59. It is characteristic that in Martin Luther's translation made at the beginning of the sixteenth century (as it was in the English Authorized Version a century later) sikera is translated as "strong drink": "Wein und stark Getränk wird er nicht trinken." The use of this phrase indicates that in Germany in 1520 the word Branntwein ("burnt", that is, distilled wine; compare "brandy"), which O'Leary, Rodes and Kilburger employed a century later to refer to Russian vodka, was not yet in use. The word Branntwein did not appear in the German language in any sense before the sixteenth century.

31. A.D. Veisman, *Grechesko-russkii slovar'*, St Petersburg 1888, p. 1127.

32. M. Fasmer, *Etimologicheskii slovar' russkogo iazyka*, vol. 3, p. 620.

33. Mikhelson, *Obiasnenie 25,000 inostrannykh slov, voshedshikh v upotreblenie v russkii iazyk*, Moscow 1865, p. 577. P. Alekseev, *Tserkovnii slovar'*, vol. 3, p. 59.

34. *Slovnik jezyka staroslovenskeho*, vol. 3, p. 35, *Nikonovskaia letopis'*.

35. Ibid., vol. 2, p. 359, *Ustiuzhskaia kormchaia*.

36. Tereshchenko, *Byt russkogo naroda*, vol. 1, St Petersburg 1848, pp. 207–8.

37. *Slovnik jezyka staroslovenskeho*, vol. 2, p. 539.

38. I. Zabelin, *Istoriia russkoi zhizni s drevneishikh vremen*, vol. 1, p. 450.

39. *Sobranie gosudarstvennykh gramot i dogovorov*, Moscow 1813, section 1, p. 2, document no. 1.

40. S.M. Solov'ev, *Istoriia Rossii s drevneishikh Vremen*, St Petersburg 1881, book 1, vol. 3, pp. 982–983.

41. *Nikonovskaia letopis'*, IV, cited in S.M. Solov'ev, book 1, vol. 4, p. 1056.

42. I. Pryzhov, *Istoriia kabakov v Rossii v sviazi s istoriei russkogo naroda*, Moscow 1868.

43. In the years from 1396 to 1528 Genoa was a political and military ally of France, which played the role of Genoa's protector. In Byzantium Genoese merchants enjoyed enormous privileges which had been granted to them by Michael Paleologus in the thirteenth century, and for this reason they also enjoyed a degree of trust in Russia, where the population was mistrustful of other foreigners.

44. A "measure" (*mera, osminnik, osmina* or *osmerik*) was equal to approximately 100 litres. A *pud* was roughly 16 kilograms. After the honey had been boiled with the hops and allowed to settle, and after the wax had been removed, about one-third of the volume remained. If we consider that the mead then began to ferment and was partly "stewed" in a stove, we can estimate that about a quarter of the original quantity remained; that is, 4 kilograms or rather less than 4 litres. When the honey was

diluted in a ratio of four to one the original volume was restored; in the process of "standing" the mead with berries the volume fell again by half. During further maturation about one-tenth of the volume was lost. It can thus be seen that for preparing a barrel of mead of forty "buckets" (about 500 litres), about ninety "buckets" of honey were required. This indicates that in the seventeenth century mead was already extremely expensive.

45. The use of isinglass in the production of mead further demonstrates that *medostav* was a process analogous to winemaking in the countries of southern Europe. Similar techniques are used to this day in the production of dessert wines, and chilling is employed as well. All these techniques were characteristic of *medostav*. A major role was also played by decanting, as in all ancient winemaking processes. This technique was so expensive as to be uneconomic, but it was effective.

46. A. Shletser (A.L. von Schlözer), *Nestor*, St Petersburg 1810, ch. ii, pp. 295, 296.

47. *Slovar' russkogo iazyka, XI–XVII vv.*, vol. 4, Moscow 1977, p. 200.

48. Ibid. See also *Akty Severo-Vostochnoi Rusi*, vol. 1, no. 607, p. 506.

49. See Nestor, *Nachalnaia letopis'; Zhitie Feodosiia; Lavrentevskaia letopis'*; I.I. Sreznevskii; *Slovar' drevne russkogo iazyka*, vol. 1, p. 1411. As this dictionary notes, the word *korchaga* is encountered in all the oldest Russian chronicles.

50. M. Fasmer, *Etimologicheskii slovar' russkogo iazyka*, vol. 2, p. 341.

51. The fact that the word *korchma*, in the sense of a tavern, an alehouse, should have its origins in the word *kuvshin* (that is, *korchaga* in the original sense) is not something unique to the Russian language. Something very similar appears in German, where the word *Krug* (jug) also has the meaning of a tavern. In Swedish the word *krog* means both a mug and an inn.

Chapter 2

1. It is significant that mead-brewing and *medostav* attained their highest level of development in Russia during the period when it was divided into small, sometimes tiny independent principalities: in the republics of Novgorod and Pskov; in the Finnic Mordva area; and also in pagan, early feudal Lithuania, which was divided into tribal territories.

2. Among these reasons are the Muslim prohibition of alcoholic liquors; in the Crimea; the tradition of winemaking and viticulture inherited from the Greeks; the availability to the Golden Horde of grape wine imported from Transcaucasia, and also the presence of vineyards in the region between the Volga and the Don; and the impossibility for nomadic and semi-nomadic peoples of establishing production of a product dependent on large-scale grain cultivation.

3. During the periods when there was no such monopoly, vodka riots occurred in Russia rather often. Thus in 1648, sixteen years after the introduction of the "tax farm" system, there occurred the greatest vodka riot in all of Russian history, which grew into the revolt of Stenka Razin. The Assembly of the Land (*Zemskii sobor*) of 1649 came out decisively for the abolition of the farm, and the "assembly on taverns" of 1652 reaffirmed this decision. In exactly the same way the vodka riots of 1859–62 were directed against the hated farm, and against the adulteration of vodka by the tax farmers who held these concessions. The riots of this period inspired various changes in the system of vodka production and in the legislation pertaining to vodka, and hastened the abolition of serfdom.

4. N.M. Karamzin gives the following description of the festive beverages provided at the time of the Troki Congress: "... every day 700 barrels of mead were brought out of the prince's cellars, apart from wine, Romanée and beer ..." The principal liquor here was mead, the main alcoholic beverage of the Lithuanians. Wine was mentioned as a sideline. We can surmise that the Romanée burgundy was being distinguished from other grape wines, but there is no doubt that it was grape wine that was being referred to. Karamzin

was extremely scrupulous, and when he used an original
source reproduced it precisely without introducing his own
interpretations. See N.M. Karamzin, *Istoriia gosudarstva
rossiiskogo*, 5th edition, Moscow 1989, vol. 5, ch. 3, col.
146.

5. *Pure* was prepared exclusively by cold fermentation and prolonged maturation; its technology resembled that of matured
mead, but without any heating at all. The production
process involved the introduction one after another of various ingredients at defined intervals, and the use of other
measures such as cooling with ice and delaying fermentation. The stages comprised a combination of malt, rye wort,
honey, extract of hops and yeast; fermentation; "ripening";
decanting; and maturation for a period of eight months to
two years.

6. The fifty-odd years between 1334 and 1386 represented the
normal period which was required at that time for a
Western European technical discovery to reach Moscow. It
took that long first for discoveries to emerge from secrecy,
and second to make their way to Moscow state. An analogy
exists in the case of gunpowder and firearms. Gunpowder,
though long known in China, was first mentioned in Europe
by Roger Bacon around 1280, and was perfected in the late
1330s. Its military uses first became known in Moscow in
1389 – again the same cycle of fifty years.

7. N.M. Karamzin, *Istoriia gosudarstva rossiiskogo*, 1st edn, St
Petersburg 1817, vol. 5, p. 46.

8. Ibid., p. 47.

9. There are rivers with such names not only in Finland and
Russian Karelia, but also in the Perm district and in the
Vyatka territories. It is also possible that something more
unusual occurred: that the river was named in commemora-
tion of a particular event without regard for its former
name.

10. The eighteenth-century historian M. Chulkov, in a work
dealing with Russian economic history and in particular
with the history of foreign trade, wrote in very definite
terms about the supposedly self-evident fact that "alcohol
distilling was brought to us from Italy by the Genoese

somewhere around the beginning of the fourteenth century". Chulkov evidently had in mind the fact, which we have noted more than once, that the Genoese brought *aqua vitae* with them to Moscow. If this is the case, Chulkov's data are erroneous; or, more precisely, they reproduce the errors of the chroniclers. But perhaps Chulkov had at his disposal factual material unknown to us today. If this were the case, however, alcohol distilling could not have arisen earlier than the second half of the 1390s. In all probability there is a simple slip of the pen here, and Chulkov was really talking about the beginning of the fifteenth, not the fourteenth century. The possibility of this cannot be rejected, since in the same book Chulkov sets out very precisely (though without citing his source) the contents of documents which became generally known only in the nineteenth century, after their publication by N.P. Rumyantsev. See M. Chulkov, *Istoriia kratkaia rossiiskoi torgovli*, Moscow 1788, p. 19). It is, however, possible that there is no slip of the pen in Chulkov's work, and that he simply reproduced the facts as they were presented to him. None of the subsequent historians of the nineteenth and twentieth centuries have had at their disposal documents or other materials establishing the presence of alcohol distilling during the fourteenth century, or even, for that matter, the fifteenth. M. Solov'ev, for example, in his multi-volume *History of Russia*, mentions vodka first in relation to the year 1558, during the reign of Ivan the Terrible.

11. We shall use the example of Sweden for purposes of comparison because, first, only Russia and Sweden developed into great powers in the area; and second, because Sweden possesses good historical statistics which make it possible to present clearly the processes for which statistical data are lacking in the case of Moscow.

12. When the state imposes a monopoly on any product, a basic consideration is that the product should have no ready replacement or surrogate. Hence salt, despite its cheapness, is an ideal product on which to levy a monopoly tax. Tea, coffee, vodka, matches and kerosene were viewed similarly in times when the state alone possessed the technical secrets of their production, or when they could not be produced by

hand with the same quality as with state equipment in state enterprises.

13. No one has remarked on this fact, since today's historians, accustomed to setting out recent events in detail and earlier ones in compressed form, find it difficult to understand any other way of proceeding. It should not be forgotten, however, that until the 1930s even Soviet historians considered it bad form to continue their record of events right up to the immediate period. Until the Revolution historians used to cut short their accounts one or two Tsars' reigns before their own time.

14. This is quite understandable. Ritual drunkenness was practised in a collective manner by large gatherings of people, while vodka (as a non-traditional drink which was, moreover, subject to a state monopoly), was consumed either in the home on an individual basis, or in taverns – that is, briefly and in small groups. In addition, distilled spirit has an antiseptic effect, so that drinking vessels would not pass on infection; while beer and ale supported the growth of numerous microorganisms.

Chapter 3

1. Winemaking in Russia is usually considered to have arisen around the beginning of the nineteenth century, following the annexation of the Crimea. But as early as 1647 a vineyard was organized on the river Terek – an official decree relating to it was issued in 1650 – and in 1659 a decision was taken to cultivate grapes and develop winemaking in Astrakhan.

2. In this case the barrel referred to was the "mead barrel" of forty "buckets". Measurements are explained later in this chapter.

3. N. Karamzin, *Istoriia gosudarstva rossiiskogo*, vol. 5, St Petersburg 1817, p. 337.

4. Ibid., vol. 6, notes, p. 467.

5. M. Chulkov, *Slovar' iuridicheskii*, Moscow 1792.

6. Ibid., pp. 282–3.

7. *Svod zakonov rossiiskoi imperii*, vol. XV, V [lozh. Nak.],

p. 1283, a prohibition on bringing beer, wine, or vodka and other "burning" (*goriachie*) drinks into the city stores and markets. A. Nurnberg, *Ukazatel'*, Moscow 1911, p. 86.

8. The Cherkassk Cossacks did not filter their vodka through charcoal because there were no birch forests in the Cherkassy region. They did not use any of the other Russian methods of purification, but masked the rancid odour with herbs and hops. This is dealt with in more detail in the section on the technology of production of vodka in the eighteenth century.

9. In Belorussia the term *raka* (or *radzia*) signified not the first stage of distillation but the second, subsequent one, known in Russia by the name *prostoe vino* (simple wine), that is, a finished product ready to be traded.

10. V. Burianov, *Progulka s det'mi po Rossii* (A stroll with children through Russia), vol. 1, St Petersburg 1837, p. 100.

11. Vol. 1, p. 151.

12. These three proverbs are: *Vodka – vinu tetka* (Vodka is the aunt of wine), *Nynche i pianitsa na vodku ne prosit, a vse na chai* (Now [things are so bad that] the drunks don't ask for vodka, but only for tea) and *Kak khochesh' zovi, tol'ko vodkoi poi* (Call me what you like, only give me vodka). The latter is a variant of an earlier proverb: *Kak khochesh' obzyvai, lish' vina nalivai* (Call me what you like, only pour out some vodka).

13. Spirituous liquors with an alcoholic content of from 65 to 70 per cent, made with the addition of sugar and herbs, began to be called "balsams", "Russian liqueurs" and "spiced brandies" (*zapekanki*), while those with a content from 70 to 75 per cent were *erofeichi*. In this way the whole scale of strength was covered.

Chapter 4

1. Oberto Spinola, *Le musée Martini. Histoire de l'oenologie*, Turin 1959, p. 35.

2. V. Prokopovich, *O rashchenii soloda*, vol. IV (XXXVI), Trudy Imperatorskogo Vol'nogo Ekonomicheskogo Obshchestva, St Petersburg 1785, p. 133.

3. P. Rychkov, *O spirte iz verbovykh tsvetov i iz travy, nazyvaemoi vorobinoi; Opyty vinnogo kureniia na domashnii raskhod; O travianykh koreshkakh i semenakh, prigodnykh k vinnoi sidke*, vol. IX, Trudy Imperatorskogo Vol'nogo Ekonomicheskogo Obshchestva, St Petersburg 1768. I.G. Model, *Sposob k vinnomu kureniiu na domashnie raskhody*, vol. IX, ibid., p. 60.

4. K.S. Kropotkin, *Vinokurenie*, St Petersburg 1889, pp. 553–4.

5. Excise is essentially an indirect tax, imposed exclusively on articles produced within the country, and affecting goods produced and sold by private individuals rather than the state. An excise differs from a monopoly in the way in which it is levied. Under the monopoly a single price was decreed for vodka, and this applied to all consumers throughout the state. Under the excise system the government imposed a tax on the private production of vodka, on the basis of payments for the possession of particular equipment, reckoned by the volume of the distilling vats; plus duties on each unit by volume of the finished product – in the nineteenth century these amounted to approximately six kopeks per bucket. The private producer set the price of vodka after calculating the cost of raw materials and processing, taxes, and projected profits from retail sale. This price fell entirely on the consumer, and varied widely in different parts of the country and in different distilling establishments. An excise can thus be described as a tax on consumption, since in the final accounting it forces the consumer to pay all the costs of the private distiller and of the state revenue collecting agency. In principle the most advantageous system for the consumer is not the farm system (renting the monopoly to private individuals), which leads to arbitrary, unjustified rises in the price of vodka. Nor is it an excise system, under which the prices for vodka are not only set arbitrarily, but also tend to be higher than when (other things being equal) the state controls the industry. The best system from the point of view of the consumer is a state monopoly, under which there is one proprietor for vodka – the treasury. Under the farm or an excise system there are at least two proprietors, or in the case of subletting, three. Each of these proprietors naturally aims to grow

rich, or at least tries not to make a loss. Most importantly, a monopoly allows strict and unified control over the quality of vodka. With the farm or an excise system such control is difficult or impossible, and there is no guarantee that the vodka will not be adulterated.

Chapter 5

1. *Russia. Official Report of the Delegation of British Trade Unionists Visiting Russia and the Caucasus in November and December 1924*, transl. from English by D. Ya. Dlugach, I.V. Rumera and A.M. Sukhotina, Moscow 1925, p. 156.
2. Ibid., pp. 148, 157.
3. The reference is to the position before 1924.
4. *Russia*, p. 156.
5. K. Marx and F. Engels, *Collected Works*, 1st Russian edn, vol. XV, Moscow 1933, p. 301.
6. In the 1950s and 1960s the American physiologist and bio-chemist R.J. Williams published a number of research papers on the biological requirements of people employed in various branches of industry. He emphasized that the need for alcohol in modern industrial society has objective biochemical causes. Williams thus confirms the observations of Friedrich Engels, using new experiments and physiological data. R.J. Williams, *Biochemical Individuality*, New York 1956. See also research by Swedish scientists: *Socialar avvikelser och socialkontrol*, Stockholm–Göteborg–Uppsala, 1964, pp. 290–91.
7. V.I. Lenin, *On the Role of the Trade Unions in Production*, text of a speech made 30 December 1920, Moscow 1921, p. 38.

Printed in the United States
by Baker & Taylor Publisher Services